Lecture Notes in Mathematics

Edited by A. Dold and B. Eckmann

Subseries: Scuola Normale Superiore, Pisa
Adviser: E. Vesentini

1022

Graziano Gentili
Simon Salamon
Jean-Pierre Vigué

Geometry Seminar "Luigi Bianchi"

Lectures given at the Scuola Normale Superiore, 1982

Edited by E. Vesentini

Springer-Verlag
Berlin Heidelberg New York Tokyo 1983

Authors

Graziano Gentili
Simon Salamon
Scuola Normale Superiore
Piazza dei Cavalieri 7, 56100 Pisa, Italy

Jean-Pierre Vigué
U.E.R. de Mathématiques, Université de Paris VI
4 Place Jussieu, 75230 Paris Cedex 05, France

Editor

Edoardo Vesentini
Scuola Normale Superiore
Piazza dei Cavalieri 7, 56100 Pisa, Italy

AMS Subject Classifications (1980): 53B, 53C, 32A, 32C, 32M, 46A, 51K

ISBN 3-540-12719-4 Springer-Verlag Berlin Heidelberg New York Tokyo
ISBN 0-387-12719-4 Springer-Verlag New York Heidelberg Berlin Tokyo

Printing and binding: Beltz Offsetdruck, Hemsbach/Bergstr.
2146/3140-543210

GRAZIANO GENTILI

DISTANCES ON CONVEX CONES

SIMON SALAMON

TOPICS IN FOUR-DIMENSIONAL RIEMANNIAN GEOMETRY

JEAN-PIERRE VIGUÉ

DOMAINES BORNÉS SYMÉTRIQUES

INTRODUCTION

The geometry seminar held during the Winter and Spring of 1982 at the Scuola Normale Superiore consisted mainly of three cycles of lectures delivered by Graziano Gentili, Simon Salamon and Jean-Pierre Vigué.

The texts published here are amplified versions of those lectures.

E.V.

C O N T E N T S

GRAZIANO GENTILI: DISTANCES ON CONVEX CONES

SIMON SALAMON: TOPICS IN FOUR-DIMENSIONAL RIEMANNIAN GEOMETRY

JEAN-PIERRE VIGUÉ : DOMAINES BORNÉS SYMÉTRIQUES

GRAZIANO GENTILI

DISTANCES ON CONVEX CONES

PREFACE

Kobayashi and Carathéodory-type pseudo-distances on convex cones in locally convex real vector spaces were defined by E. Vesentini [7]. Their most characteristic property is the following:

P) Every linear homomorphism between any two cones is a contraction with respect to the Kobayashi (Carathéodory) pseudo-distance.

In this work all the distances (on cones of dimension greater than one) having property P) will be classified.

For definitions and properties of Kobayashi and Carathéodory-type pseudo-distances (denoted respectively by γ and δ) we refer to Vesentini [7] and Franzoni [1].

PRELIMINARIES AND FORMULATION OF THE PROBLEM

Let R be a real vector space and let $V \neq \{0\}$ be a convex cone in R, i.e. a subset of R such that $v \in V \Rightarrow tv \in V$ for all $t \in \mathbb{R}^+_*$, and that $u,v \in V \Rightarrow u + v \in V$. We shall always suppose that R coincides with the vector space $Sp(V) = V - V$ spanned by V, and shall be concerned mainly with the case in which V is <u>sharp</u> (i.e. does not contain any entire affine straight line of R) and satisfies the following condition:

i) If $v \in V$ and if $\mathcal{L}$ is any affine straight line in R such that $v \in \mathcal{L}$ and that $\mathcal{L} \cap V$ contains a half-line, then there is a half-line $\mathcal{R}_v \subset \mathcal{L} \cap V$ containing v in its interior.

Condition i) turns out to be equivalent to the following (see Vesentini [7]):

i)' For every finite-dimensional subspace S of R such that $S \cap V \neq \{0\}$, $S \cap V$ is open in the subspace $Sp(S \cap V)$ of S generated by $S \cap V$, with respect to the Euclidean topology.

For the sharp convex cone $\mathbb{R}^+_*$ of $\mathbb{R}$ we shall take as a distance between any two points x and y the Haar measure (with respect to the multiplicative group $\mathbb{R}^+_*$) of the interval determined by x and y:

$$\sigma(x,y) = \left|\log \frac{x}{y}\right|.$$

The distance σ is invariant with respect to the action of the group

$$GL(\mathbb{R}^+_*) = \{f \in GL(\mathbb{R}) : f(\mathbb{R}^+_*) = \mathbb{R}^+_*\};$$

moreover for every affine endomorphism g of $\mathbb{R}$ such that $g(\mathbb{R}^+_*) \subset \mathbb{R}^+_*$ we have

$$\sigma(g(x),g(y)) \leq \sigma(x,y) \qquad (x,y \in \mathbb{R}^+_\star).$$

By means of the distance σ on $\mathbb{R}^+_\star$, the Kobayashi-type and Carathéodory-type pseudo-distances can be defined on any convex cone V satisfying condition i) (see Vesentini [7]).

The Kobayashi-type pseudo-distance.

Let $w_0 = u, w_1, \ldots, w_n = v$ be points in V, let $x_1, \ldots, x_n$, $y_1, \ldots, y_n$ be points in $\mathbb{R}^+_\star$ and let $f_1, \ldots, f_n$ be affine functions of $\mathbb{R}$ into R mapping $\mathbb{R}^+_\star$ into V, and such that

$$f_j(x_j) = w_{j-1}, \; f_j(y_j) = w_j \qquad (j = 1, \ldots, n).$$

Let us define

$$\gamma_V(u,v) = \inf\{\sigma(x_1,y_1) + \ldots + \sigma(x_n,y_n)\}$$

taking the infimum over all possible choices of $n, w_1, \ldots, w_{n-1}$, $x_1, \ldots, x_n$, $y_1, \ldots, y_n$, $f_1, \ldots, f_n$. The function γ_V is a pseudo-distance for every V and property P) (see the Preface) is fulfilled. Moreover it turns out that:

K1. The pseudo-distance γ_V is a distance if, and only if, V is sharp.

K2. If $R = \mathbb{R} \times \mathbb{R}$ and $V = \mathbb{R}^+_\star \times \mathbb{R}^+_\star$ then

$$\gamma_{\mathbb{R}^+_\star \times \mathbb{R}^+_\star}((x_1,x_2),(y_1,y_2)) = \max\,(\sigma(x_1,y_1),\sigma(x_2,y_2))$$

or

$$\gamma_{\mathbb{R}^+_\star \times \mathbb{R}^+_\star}((x_1,x_2),(y_1,y_2)) = \sigma(x_1,y_1) + \sigma(x_2,y_2),$$

depending on whether the affine line determined by (x_1,x_2) and (y_1,y_2) intersects $\mathbb{R}^+_\star \times \mathbb{R}^+_\star$ in a half-line or in a segment.

The Carathéodory-type pseudo-distance.

Let $F(V)$ be the collection of all real-valued affine functionals on R mapping V into $\mathbb{R}^+_\star$. Let $u,v \in V$ and let

$$\delta_V(u,v) = \sup\{\sigma(f(u),f(v)) \;:\; f \in F(V)\}.$$

The function δ_V is a pseudo-distance for any V satisfying condition i), and property P) (see the Preface) is fulfilled. Moreover:

C1. The pseudo-distance δ_V is a distance if, and only if, V is sharp.

C2. If $R = \mathbb{R} \times \mathbb{R}$ and $V = \mathbb{R}^+_\star \times \mathbb{R}^+_\star$ then

$$\delta_{\mathbb{R}^+_\star \times \mathbb{R}^+_\star}((x_1,x_2),(y_1,y_2)) = \max\{\sigma(x_1,y_1),\sigma(x_2,y_2)\}.$$

C3. $\delta_V(u,v) = \sup\{\sigma(\lambda(u),\lambda(v))\}$,

where the supremum is taken over all linear forms λ on R mapping V into $\mathbb{R}^+_\star$.

<u>Formulation of the problem.</u>

Given two convex cones $V_1 \subset R_1$ and $V_2 \subset R_2$ of two real vector spaces R_1 and R_2, we shall denote by $\mathrm{End}(V_1,V_2)$ the semigroup of all linear maps ϕ of R_1 into R_2 such that $\phi(V_1) \subset V_2$.
Let $d^1_{V_1}$ and $d^2_{V_2}$ be two distances defined on V_1 and V_2 respectively.

If

$$d^2_{V_2}(\phi(u),\phi(v)) \leq d^1_{V_1}(u,v)$$

for all $u,v \in V_1$ and all $\phi \in \mathrm{End}(V_1,V_2)$ and if

$$d^1_{V_1}(\psi(p),\psi(q)) \leq d^2_{V_2}(p,q)$$

for all $p,q \in V_2$ and all $\psi \in \mathrm{End}(V_2,V_1)$ then the pair of distances $(d^1_{V_1},d^2_{V_2})$ is called a <u>special pair (of distances)</u> on (V_1,V_2).

When $R_1 = R_2 = R$ and $V_1 = V_2 = V$, it follows that $d^1_{V_1} = d^2_{V_2} = d_V$. In this situation d_V is called a <u>special distance</u> on V.

Special pairs of distances and special distances are obviously invariant with respect to the action of all linear isomorphisms between the cones in question.

Examples:

1. If V_1 and V_2 are convex sharp cones satisfying property i), then $(\gamma_{V_1}, \gamma_{V_2})$ and $(\delta_{V_1}, \delta_{V_2})$ are special pairs on (V_1, V_2).

2. If V is a convex sharp cone satisfying property i), then γ_V and δ_V are special distances on V.

The following questions arise:

a) Given two cones V_1 and V_2, how many different special pairs of distances exist on (V_1, V_2)?

b) Given a cone V, how many different special distances can be defined on V?

c) Are all these distances "essentially planar"? In other words, do they depend only on the two-dimensional subspace determined by the two points which we calculate the distance between?

In this work we give a characterization of all the special pairs of distances on any two convex sharp cones, V_1 and V_2, satisfying property i) and having dimensions strictly greater than 1. We shall at first relate the Carathéodory distance on V with the order V induces on the spanned vector space $Sp(V) = R$; this will give us the main tool to obtain a classification.

1. ORDER AND CARATHEODORY PSEUDO-DISTANCE

Let V be a convex cone not containing 0, in the vector space $R = Sp(V)$. By means of the cone V we define on R the relation $<$ in the following manner:

Definition 1.1 For $u,v \in R$, $u < v$ if $v - u \in V$.

Proposition 1.2 The following properties hold:

R1. $u < v$ and $v < w$ imply $u < w$,

R2. $u < v$ if and only if $u + w < v + w$,

R3. $u < v$ if and only if $tu < tv$,

for all $u,v,w \in R$ and all $t \in \mathbb{R}_\star^+$. Moreover:

R4. $t < s \Rightarrow tu < su$,

for all $u \in V$ and all $t,s \in \mathbb{R}_\star^+$.

The relation $<$ is transitive and strictly antisymmetric, i.e. is a strict order.

Proof.

R1. $u < v \Leftrightarrow v-u \in V$, and $v < w \Leftrightarrow w-v \in V$. V being convex, we obtain $(v-u) + (w-v) = w-u \in V$, whence $u < w$.

R2. $v-u = (v+w) - (u+w) \in V$.

R3. $v-u \in V \Leftrightarrow t(v-u) \in V$ for all $t \in \mathbb{R}_\star^+$.

R4. $u \in V$ and $(s-t) \in \mathbb{R}_\star^+ \Rightarrow (s-t)u \in V$.

Antisymmetry: suppose that $u < v$ and $v < u$ $(u,v \in R)$; this means that $v-u, u-v \in V$ yielding $(v-u) + (u-v) = 0 \in V$, which contradicts the assumptions. ∎

If $W \subset R$ is another convex cone and if $W \subset V$, then W is called a subcone of V.

<u>Lemma 1.3</u> Let W be a subcone of V. $W = Sp(W) \cap V$ if and only if the order $<_W$ defined on $Sp(W)$ by W according to Definition 1.1 coincides with the restriction to $Sp(W)$ of $<_V$.

<u>Proof</u>.
Let $u,v \in Sp(W)$ be such that

$$u <_W v.$$

Then $v-u \in W \subset V$, yielding

$$u <_V v.$$

On the other hand, if $u,v \in Sp(W)$ are such that

$$u <_V v$$

then $v-u \in V \cap Sp(W) = W$, i.e.

$$u <_W v.$$

Conversely, if $W \subsetneq Sp(W) \cap V$ there exists $v_0 \in Sp(W) \cap V$ such that $v_0 \notin W$; it follows that $0 <_V v_0$ and $0 \not<_W v_0$. ∎

<u>Theorem 1.4</u> The convex cone V (not containing 0) satisfies property i) if, and only if, for all $u,v \in V$ the set

$$\{t \in \mathbb{R}^+_\star : u < tv\}$$

is non-empty.

<u>Proof</u>.
If $u,v \in V$ and if $\Pi(u,v)$ is the (one or two-dimensional) subspace of R spanned by u and v, then, for $t \in \mathbb{R}^+_\star$

$$u < tv \iff tv-u \in V \cap \Pi(u,v) \iff v - \frac{1}{t}u \in V \cap \Pi(u,v).$$

Hence if condition i) (equivalent to i)', see the Preliminaries) is satisfied by V, $V \cap \Pi(u,v)$ is open with respect to the Euclidean topology of $\Pi(u,v)$ and then there exists $t \in \mathbb{R}^+_\star$ such that $u < tv$. To prove the converse, notice that (since $Sp(V) = R$) for all $r \in R$ there exist $w,v \in V$ such that $r = w-v$; hence for every $u \in V$ and all $t \in \mathbb{R}^+_\star$

$$tu+r = (tu-v) + w.$$

By hypothesis there exists $t_0 \in \mathbb{R}^+_\star$ such that

$$(t_0u-v) + w = t_0u+r \in V$$

and that

$$u + \frac{r}{t_0} \in V$$

giving the radiality of V at u, which implies the assertion. ∎

As a spin-off of the preceding proof we obtain:

<u>Corollary 1.5</u> The convex cone V (not containing 0) satisfies property i) if, and only if, it is radial at every point.

The hypothesis that V does not contain 0, mentioned in Theorem 1.4 and Corollary 1.5, even if not necessary for the validity of the statements, is quite reasonable, as the following proposition clarifies:

<u>Proposition 1.6</u> A convex cone V containing 0 and satisfying property i) coincides with R.

<u>Proof</u>.
Because $R = Sp(V) = V-V$, it is enough to prove that $-V \subset V$. Let $u \in V$ and let $\mathcal{L}$ be the straight line $\{tu : t \in \mathbb{R}\}$. By property i) there exists a half-line $\mathcal{R}_0 \subset \mathcal{L} \cap V$ containing 0 in its interior, i.e. there exists $s \in \mathbb{R}^+_\star$ such that $-su \in \mathcal{R}_0$. Therefore $-u \in V$. ∎

The fact that a subset of R is radial at every point has a precise topological significance: it means that the subset is open with respect to a suitable topology of R, the so-called finite topology. (See, for example [2]). From now on, R will be endowed with the finite topology.

Theorem 1.4 allows us to define the following function on any open convex cone $V \subset R$ such that $0 \notin V$:

$$K_V : V \times V \longrightarrow \mathbb{R}^+$$

$$K_V(u,v) = \inf\{t \in \mathbb{R}^+_\star : u < tv\}.$$

If $u,v \in V$, by Proposition 1.2

$$\{t \in \mathbb{R}^+_\star : u < tv\}\cdot\{s \in \mathbb{R}^+_\star : v < su\} \subset \{p \in \mathbb{R}^+_\star : u < pu\}$$

implying

(1) $$K_V(u,v)K_V(v,u) \geq K_V(u,u) = 1.$$

Hence $K_V(u,v) \in \mathbb{R}^+_\star$ and the following relation holds, for $u,v,w \in V$:

(2) $$K_V(u,v)\cdot K_V(v,w) \geq K_V(u,w).$$

Theorem 1.7 If $V_1 \subset R_1$ and $V_2 \subset R_2$ are two convex cones, open and without 0, then

$$K_{V_2}(\phi(u),\phi(v)) \leq K_{V_1}(u,v)$$

for all $\phi \in \mathrm{End}(V_1,V_2)$ and all $u,v \in V_1$. In particular, if $V_1 = V_2 = V$ and $R_1 = R_2 = R$, K_V is invariant with respect to the action of the group of all linear automorphisms of V.

Proof.

Let $t \in \mathbb{R}^+_\star$ and $u,v \in V_1$ be such that $tv-u \in V_1$, then $\phi(tv-u) = t\phi(v) - \phi(u) \in V_2$. ∎

Relation (1) implies that

$$\max\{K_V(u,v), K_V(v,u)\} \geq 1$$

for all $u,v \in V$. Therefore the following function is well-defined:

$$\eta_V : V \times V \longrightarrow \mathbb{R}^+$$

$$\eta_V(u,v) = \max\{\log K_V(u,v), \log K_V(v,u)\} .$$

Theorem 1.8 The function η_V is a pseudo-distance on V which coincides with the Carathéodory-type pseudo-distance δ_V. In particular η_V is a distance if, and only if, V is sharp.

Proof.
Let $u,v \in V$ and let $f \in \mathrm{End}(V,\mathbb{R}^+_\star)$ be such that $f(v) \geq f(u)$. Hence

$$\frac{f(v)}{f(u)} \cdot u-v \notin V$$

because we have

$$f\left(\frac{f(v)}{f(u)} \cdot u-v\right) = f(v)-f(v) = 0 \notin \mathbb{R}^+_\star .$$

Therefore

$$\frac{f(v)}{f(u)} \leq K_V(v,u) \leq \max\{K_V(v,u), K_V(u,v)\}$$

giving (see property C3 of the Preliminaries):

$$\delta_V(u,v) \leq \eta_V(u,v). \tag{3}$$

As a consequence of the Hahn-Banach Theorem, for any $v_0 \in R \setminus V$ there exists $F \in \mathrm{End}(V,\mathbb{R}^+_\star)$ such that $F(v_0) \leq 0 < F(v)$, for all $v \in V$. Let $\alpha = \max\{K_V(u,v), K_V(v,u)\}$ and suppose, for example, that $\alpha = K_V(u,v) \geq 1$. By definition, for all $\varepsilon > 0$

$$(\alpha-\varepsilon)v-u \notin V.$$

Hence there exists $F_\varepsilon \in \mathrm{End}(V,\mathbb{R}^+_\star)$ such that

$$F_\varepsilon((\alpha-\varepsilon)v-u) \leq 0, \; F_\varepsilon(u) > 0, \; F_\varepsilon(v) > 0$$

yielding

$$(\alpha-\varepsilon)F_\varepsilon(v)-F_\varepsilon(u) \leq 0$$

$$\frac{F_\varepsilon(u)}{F_\varepsilon(v)} \geq \alpha-\varepsilon .$$

In conclusion

$$\delta_V(u,v) \geq \eta_V(u,v)$$

which, together with (3), proves the assertion. ∎

2. SPECIAL DISTANCES ON $\mathbb{R}_\star^+ \times \mathbb{R}_\star^+$

Let d be a special distance on the set $A = \mathbb{R}_\star^+ \times \mathbb{R}_\star^+$, i.e. a distance such that:

$$d(f(u),f(v)) \le d(u,v)$$

for all $u,v \in A$ and all $f \in \mathrm{End}(A)$. Let us define the function

$$D_d : A \longrightarrow \mathbb{R}^+$$

$$(x,y) \longmapsto d((1,1),(x,y)).$$

<u>Lemma 2.1</u> The function D_d has the following properties:

D1. $D_d(x,y) = 0 \iff (x,y) = (1,1)$,

D2. $D_d(x,y) \le D_d(s,t) + D_d(\frac{x}{s},\frac{y}{t})$,

D3. $D_d(x,y) = D_d(\frac{1}{x},\frac{1}{y})$,

D4. $D_d(f(x,y)) \le D_d(x,y)$,

for all $x,y,s,t \in \mathbb{R}_\star^+$ and all $f \in G = \{f \in \mathrm{End}(A) : f(1,1) = (1,1)\}$.

<u>Proof</u>.

D2. $D_d(x,y) = d((1,1),(x,y)) \le d((1,1),(s,t)) + d((s,t),(x,y)) =$

$= d((1,1),(s,t)) + d((1,1),(\frac{x}{s},\frac{y}{t})) = D_d(s,t) + D_d(\frac{x}{s},\frac{y}{t})$.

D3. $D_d(x,y) = d((1,1),(x,y)) = d((\frac{1}{x},\frac{1}{y}),(1,1)) =$

$= d((1,1),(\frac{1}{x},\frac{1}{y})) = D_d(\frac{1}{x},\frac{1}{y})$.

D4. $D_d(f(x,y)) = d((1,1),f(x,y)) = d(f(1,1),f(x,y)) \le$

$\le d((1,1),(x,y)) = D_d(x,y)$. ■

The function D_d completely determines the special distance d, as the following Lemma states.

Lemma 2.2 Let $D : A \longrightarrow \mathbb{R}^+$ be a function having properties D1, D2, D3, D4 listed above. Then there exists one, and only one, special distance d_D on A such that $d_D((1,1),(x,y)) = D(x,y)$ for all $(x,y) \in A$.

Proof.
Let us define

$$d_D((s,t),(x,y)) = D(\frac{x}{s},\frac{y}{t})$$

for all $s,t,x,y \in \mathbb{R}^+_\star$. Then:

1. $d_D((s,t),(x,y)) = 0$ if and only if $D(\frac{x}{s},\frac{y}{t}) = 0$ which, by D1, implies $(\frac{x}{s},\frac{y}{t}) = (1,1)$ i.e. $(s,t) = (x,y)$.

2. Property D3 gives

$$D(\frac{x}{s},\frac{y}{t}) = D(\frac{s}{x},\frac{t}{y}),$$

whence the symmetry of d_D.

3. By means of D2 we obtain

$$D(\frac{x}{s},\frac{y}{t}) \leq D(\frac{p}{s},\frac{q}{t}) + D(\frac{x}{p},\frac{y}{q})$$

(for all $p,q \in \mathbb{R}^+_\star$) which is equivalent to the triangular property for d_D.

Hence d_D is a distance. Actually d_D is a special distance. In fact, let $(s,t),(x,y) \in A$ and let g be any element of $\mathrm{End}(A)$. If $h \in GL(A)$ is the map

$$(p,q) \longmapsto (ps,qt)$$

and if $k \in GL(A)$ is such that

$$k(g(s,t)) = (1,1),$$

it follows that

$$d_D(g(s,t),g(x,y)) = d_D(g\circ h(1,1),g\circ h(\tfrac{x}{s},\tfrac{y}{t})) = D(k\circ g\circ h(\tfrac{x}{s},\tfrac{y}{t})).$$

Because $k\circ g\circ h \in G$, property D4 yields

$$d_D(g(s,t),g(x,y)) = D(k\circ g\circ h(\tfrac{x}{s},\tfrac{y}{t})) \leq D(\tfrac{x}{s},\tfrac{y}{t}) = d_D((s,t),(x,y)).$$

Moreover, if d' is a second distance with the required properties, then

$$d'((s,t),(x,y)) = d'((1,1),(\tfrac{x}{s},\tfrac{y}{t})) = D(\tfrac{x}{s},\tfrac{y}{t}) =$$
$$= d_D((1,1),(\tfrac{x}{s},\tfrac{y}{t})) = d_D((s,t),(x,y)),$$

for all $(s,t),(x,y) \in A$. ■

For $(x,y) \in A$, let $\Gamma(x,y)$ be the orbit of (x,y) with respect to the action of the semigroup

$$G = \{f \in End(A) : f(1,1) = (1,1)\}.$$

If $f \in G$ is defined by

$$f(s,t) = \begin{pmatrix} \alpha & \beta \\ \gamma & \delta \end{pmatrix}\begin{pmatrix} s \\ t \end{pmatrix} \qquad (\alpha,\beta,\gamma,\delta \in \mathbb{R})$$

then $\alpha,\beta,\gamma,\delta \geq 0$ and

$$\begin{cases} \alpha = 1-\beta \\ \gamma = 1-\delta. \end{cases}$$

Thus

$$\Gamma(x,y) = \{((1-\beta)x+\beta y,(1-\delta)x+\delta y) : 0 \leq \beta \leq 1,\ 0 \leq \delta \leq 1\},$$

i.e. the orbit $\Gamma(x,y)$ fills the square of vertices $(x,y),(x,x),(y,x),(y,y)$.

Let $D : A \longrightarrow \mathbb{R}^+$ be any function, and for $a \in \mathbb{R}^+$ let

$$B_D(a) = \{(x,y) \in A : D(x,y) \leq a\}.$$

Then the following Lemma is an immediate consequence of the above considerations:

Lemma 2.3 Given any function $D : A \longrightarrow \mathbb{R}^+$, the two following conditions are equivalent:

D4. $D(f(x,y)) \leq D(x,y)$, for all $(x,y) \in A$ and all $f \in G$;

D4'. $(x,y) \in B_D(\alpha) \Rightarrow \Gamma(x,y) \subset B_D(\alpha)$, for all $(x,y) \in A$ and all $\alpha \in \mathbb{R}^+$.

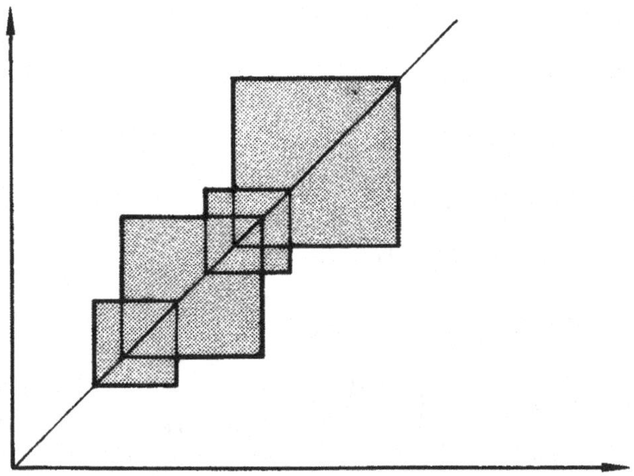

Property D4'

Now, let S be the set of all functions $g : \mathbb{R}^2 \longrightarrow \mathbb{R}^+$ satisfying the following properties for all $a,b,c,d \in \mathbb{R}$ and all $\alpha \in R^+$:

S1. $g(a,b) = 0 \Leftrightarrow (a,b) = (0,0)$,

S2. $g(a+c,b+d) \leq g(a,b) + g(c,d)$,

S3. $g(a,b) = g(-a,-b)$,

S4. $(a,b) \in B_g(\alpha) \Rightarrow \Gamma(a,b) \subset B_g(\alpha)$,
where $B_g(\alpha) = \{(p,q) \in \mathbb{R}^2 : g(p,q) \leq \alpha\}$.

The elements of S will be called <u>special functions</u>.

Let us define for any function $D : A \longrightarrow \mathbb{R}^+$ the function $D \circ \exp$ by

$$D \circ \exp(a,b) = D(e^a,e^b),$$

for all $(a,b) \in \mathbb{R}^2$.

Theorem 2.4 Let $d : A \times A \longrightarrow \mathbb{R}^+$ be a distance. Then d is a special distance if, and only if, there exists $g \in S$ such that

$$g(a,b) = d((1,1),(e^a,e^b)),$$

for all $(a,b) \in \mathbb{R}^2$.

Proof.
When d is a special distance, the function $D_d(x,y) = d((1,1),(x,y))$ satisfies properties D1, D2, D3, D4' (see Lemmas 2.1 and 2.3), so the function

$$g : \mathbb{R}^2 \longrightarrow \mathbb{R}^+$$

defined by

$$(a,b) \longmapsto D_d(e^a,e^b)$$

obviously verifies S1, S2, S3. Furthermore

$$(a,b) \in B_g(\alpha) \iff (e^a,e^b) \in B_{D_d}(\alpha)$$

giving (see D4')

$$\Gamma(e^a,e^b) \subset B_{D_d}(\alpha)$$

which, on the other hand, is equivalent to

$$\Gamma(a,b) \subset B_g(\alpha).$$

Thus g also verifies the property S4. The converse can be proved using similar arguments. ∎

As a consequence of Theorem 2.4 a large class of special distances on A can be constructed.

For every $s \in \mathbb{R}^+_\star$, let Q_s be the subset of $\mathbb{R}^2$ defined by:

$$Q_s = \{(a,b) \in \mathbb{R}^2 : b \geq a , b \geq -a , b \leq a+2s\}.$$

For every $r \geq s$ and every function $f : [-s,r] \longrightarrow \mathbb{R}^+$ such that:

1) $f(-s) = s$ and $f(r) = r$,
2) f is concave,
3) f is not decreasing,
4) the graph of f is contained in Q_s,

we set

$$Q_f = \{(a,b) \in Q_s : b \leq f(a)\}$$

and

$$B_f = \{(a,b) \in \mathbb{R}^2 : (a,b) \text{ or } (b,a) \text{ or } (-a,-b) \text{ or } (-b,-a) \text{ belongs to } Q_f\}.$$

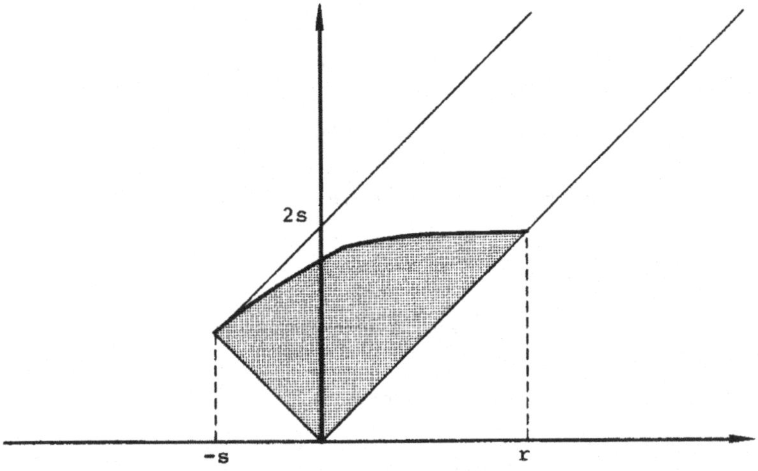

Example of a set Q_f

Hence:

<u>Proposition 2.5</u> The set B_f defined above is convex and its Minkowski functional

$$\mu_f(p) = \inf\{t \in \mathbb{R}^+ : p \in tB_f\}$$

is a special function (i.e. $\mu_f \in S$).

<u>Proof</u>.
Property S4 of μ_f is implied by property 3) of the function f. Subadditivity S2 follows from the fact that B_f is convex, and property S3 from the fact that B_f is symmetric with respect to the origin of $\mathbb{R}^2$. ∎

Given any $r \geq s > 0$ and any function $f : [-s,r] \longrightarrow \mathbb{R}^+$ satisfying properties 1), 2), 3), 4) we obtain, by means of Theorem 2.4, a special distance on A by setting

$$(4) \qquad d_f((1,1),(x,y)) = \mu_f(\log x, \log y),$$

for all $(x,y) \in A$.

Examples.

From formula C2 given in the Preliminaries we get, for the Carathéodory-type distance of A:

$$\delta_A((1,1),(x,y)) = \max(|\log x|, |\log y|).$$

Therefore Theorem 2.4 implies that the special function $g \in S$ associated to δ_A is defined by

$$g(a,b) = \max(|a|,|b|),$$

for all $(a,b) \in \mathbb{R}^2$. This $g \in S$ is the Minkowski functional of the ball $B_f \subset \mathbb{R}^2$ associated to the function $f(x) = 1$ (defined on the interval $[-1,1]$) by means of the above construction.

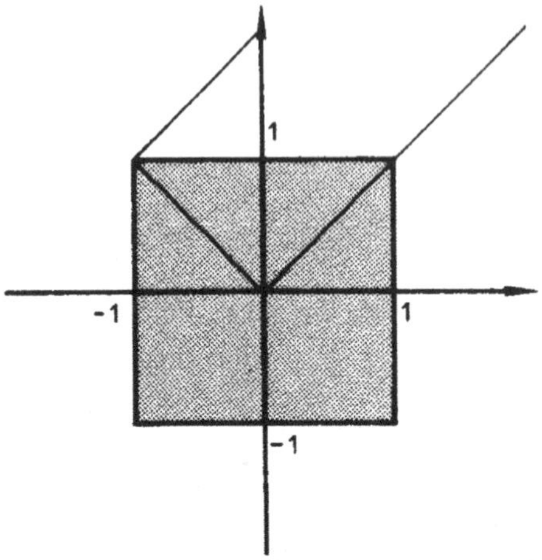

The ball $B_f \subset \mathbb{R}^2$

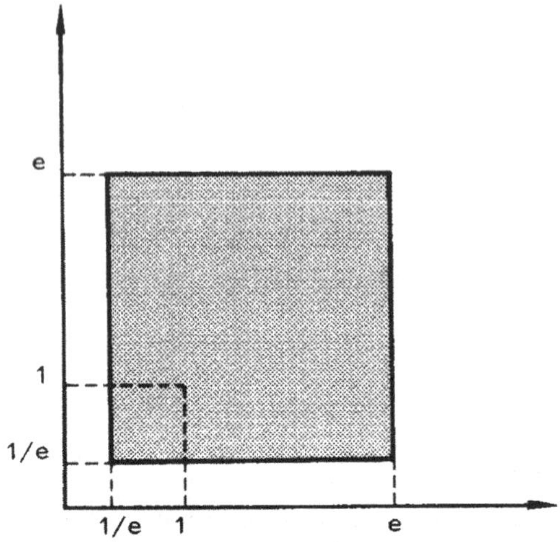

The unit ball with center (1,1) for δ_A

Analogously, for the Kobayashi-type distance on A formula K2 of the Preliminaries gives

$$\gamma_A((1,1),(x,y)) = \max(|\log x|, |\log y|)$$

or

$$\gamma_A((1,1),(x,y)) = |\log x| + |\log y|$$

according as $\log x$ and $\log y$ have the same sign or not. Again Theorem 2.4 implies that the special function $h \in S$ associated to γ_A is defined by

$$h(a,b) = \max(|a|, |b|)$$

or

$$h(a,b) = |a|+|b| \qquad ((a,b) \in \mathbb{R}^2)$$

depending on whether a and b have the same sign or not. The special function h is the Minkowski functional of the ball $B_k \subset \mathbb{R}^2$ constructed using the function

$$k : [-\tfrac{1}{2}, 1] \longrightarrow \mathbb{R}^+$$

defined by

$$k(x) = \begin{cases} x+1 & \text{for } -\frac{1}{2} \leq x \leq 0 \\ 1 & \text{for } 0 \leq x \leq 1. \end{cases}$$

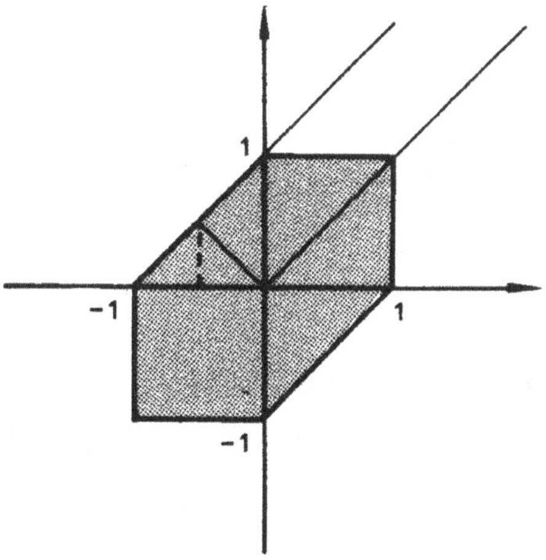

The ball $B_k \subset \mathbb{R}^2$

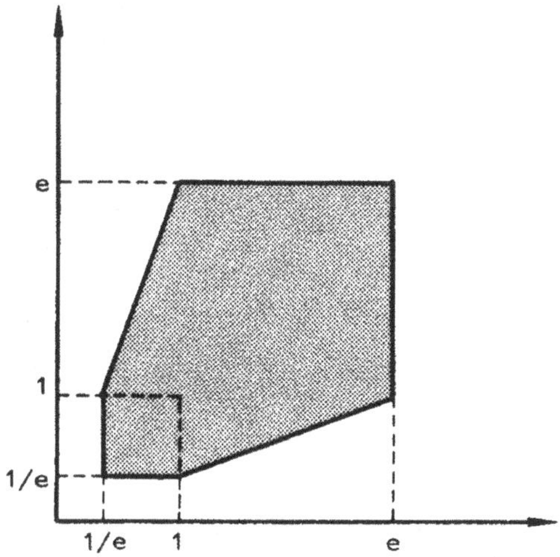

The unit ball with center (1,1) for γ_A

3. CLASSIFICATION OF ALL SPECIAL PAIRS OF DISTANCES IN DIMENSION GREATER THAN ONE

Let R, R_1, R_2 be real vector spaces, endowed with the finite topology (see, for example [2]).

<u>Theorem 3.1</u> Let $g \in S$ be a special function:

a) If V is an open, convex, sharp cone in R then the function

$$d_V^g : V \times V \longrightarrow \mathbb{R}^+$$

defined by

$$d_V^g(u,v) = g(-\log K_V(u,v), \log K_V(v,u))$$

is a special distance on V.

b) If V_1 and V_2 are open, convex, sharp cones in R_1 and R_2 respectively, then $(d_{V_1}^g, d_{V_2}^g)$ is a special pair of distances on (V_1, V_2).

<u>Proof</u>.

For $u,v \in V$, $d_V^g(u,v) = 0$ if and only if $g(-\log K_V(u,v), \log K_V(v,u)) = 0$ which is equivalent (see S1, section 2) to $\log K_V(u,v) = \log K_V(v,u) = 0$, i.e. to

$$K_V(u,v) = K_V(v,u) = 1. \tag{5}$$

Equalities (5) imply, by definition of K_V, that for all $\varepsilon, \delta > 0$

$$(\varepsilon+1)v-u \in V \quad \text{and} \quad (\delta+1)u-v \in V. \tag{6}$$

Since V is open for the finite topology of R, the cone $V \cap \Pi(u,v)$ ($\Pi(u,v)$ is the subspace spanned by u and v) is open in the finite dimensional vector space $\Pi(u,v)$. Therefore relations (6) yield

$$v-u \quad \text{and} \quad u-v \in \overline{V \cap \Pi(u,v)}.$$

The cone $\overline{V \cap \Pi(u,v)}$ being sharp, we obtain $v-u = 0$, i.e. $v = u$.

(Symmetry). Making use of the properties S3, S4 of g (see section 2) we get

$$\begin{aligned} d_V^g(u,v) &= g(-\log K_V(u,v), \log K_V(v,u)) = \\ &= g(\log K_V(v,u), -\log K_V(u,v)) = \\ &= g(-\log K_V(v,u), \log K_V(u,v)) = d_V^g(v,u). \end{aligned}$$

(Triangular property). Relation (1) of section 1 gives

$$\log K_V(v,u) + \log K_V(u,v) \geq 0$$

i.e.

$$(7) \qquad -\log K_V(u,v) \leq \log K_V(v,u),$$

for all $u,v \in V$.

On the other hand, by (2) of section 1:

$$\log K_V(u,v) \leq \log(K_V(u,w)K_V(w,v)),$$

for all $u,v,w \in V$. Therefore

$$(-\log K_V(u,v), \log K_V(v,u)) \in \Gamma(-\log(K_V(u,w)K_V(w,v)), \log(K_V(v,w)K_V(w,u)))$$

which used together with the subadditivity of g gives:

$$\begin{aligned} d_V^g(u,v) &= g(-\log K_V(u,v), \log K_V(v,u)) \leq \\ &\leq g(-\log(K_V(u,w)K_V(w,v)), \log(K_V(v,w)K_V(w,u))) = \\ &= g(-\log K_V(u,w) - \log K_V(w,v), \log K_V(v,w) + \log K_V(w,u)) \leq \\ &\leq g(-\log K_V(u,w), \log K_V(w,u)) + g(-\log K_V(w,v), \log K_V(v,w)) = \\ &= d_V^g(u,w) + d_V^g(w,v). \end{aligned}$$

This proves that d_V^g is a distance. Furthermore Theorem 1.7 implies that

(8) $$\log K_{V_2}(\phi(u),\phi(v)) \leq \log K_{V_1}(u,v),$$

for all $u,v \in V_1$ and all $\phi \in \mathrm{End}(V_1,V_2)$. Relations (7) and (8) give

$$(-\log K_{V_2}(\phi(u),\phi(v)),\log K_{V_2}(\phi(v),\phi(u))) \in \Gamma(-\log K_{V_1}(u,v),\log K_{V_1}(v,u))$$

whence

$$d^g_{V_2}(\phi(u),\phi(v)) = g(-\log K_{V_2}(\phi(u),\phi(v)),\log K_{V_2}(\phi(v),\phi(u))) \leq$$
$$\leq g(-\log K_{V_1}(u,v),\log K_{V_1}(v,u)) = d^g_{V_1}(u,v)$$

which completes the proof of parts a) and b). ∎

The following Lemma joins together the results of Theorems 2.4 and 3.1, in the case of the cone $\mathbb{R}^+_\star \times \mathbb{R}^+_\star = A$.

<u>Lemma 3.2</u> Let $d : A \times A \longrightarrow \mathbb{R}^+$ be a special distance and let $g \in S$ be the special function

$$g(a,b) = d((1,1),(e^a,e^b))$$

(for $(a,b) \in \mathbb{R}^2$). Then

$$d = d^g_A .$$

<u>Proof</u>.

Let us point out that, if $x = (x_1,x_2)$ and $y = (y_1,y_2)$ belong to A, we can suppose

$$\frac{x_1}{y_1} \leq \frac{x_2}{y_2} .$$

In this situation

$$K_A(x,y) = \frac{x_2}{y_2}$$
$$K_A(y,x) = \frac{y_1}{x_1}$$

and

$$
\begin{aligned}
d(y,x) &= d((y_1,y_2),(x_1,x_2)) = \\
&= d((1,1),(\frac{x_1}{y_1},\frac{x_2}{y_2})) = \\
&= g(\log \frac{x_1}{y_1}, \log \frac{x_2}{y_2}) = \\
&= g(-\log \frac{y_1}{x_1}, \log \frac{x_2}{y_2}) = \\
&= g(-\log K_A(y,x), \log K_A(x,y)) = d_A^g(y,x).
\end{aligned}
$$

Since $x,y \in A$ are arbitrary, the assertion follows. ∎

<u>Lemma 3.3</u> Let Q_1 and Q_2 be cones in $\mathbb{R}^2$ isomorphic to $A = \mathbb{R}_\star^+ \times \mathbb{R}_\star^+$. If (d_1,d_2) is a special pair of distances on (Q_1,Q_2), then there exists one and only one special function $h \in S$ such that

$$d_1 = d_{Q_1}^h$$

and that

$$d_2 = d_{Q_2}^h .$$

<u>Proof</u>.

Let $\phi : A \longrightarrow Q_1$ and $\psi : A \longrightarrow Q_2$ be two linear isomorphisms, and let $\tilde{d}_1$ and $\tilde{d}_2$ be the two distances on A defined by

$$\tilde{d}_1(x,y) = d_1(\phi(x),\phi(y))$$

and

$$\tilde{d}_2(x,y) = d_2(\psi(x),\psi(y)) \qquad (x,y \in A).$$

$$
\begin{array}{ccc}
Q_1 & & Q_2 \\
\phi \big\uparrow & & \big\uparrow \psi \\
A & \xrightarrow[\eta]{\qquad} & A
\end{array}
$$

For every $\eta \in \mathrm{End}(A)$ the map $\psi \circ \eta \circ \phi^{-1}$ is a linear map of Q_1 into Q_2. Hence for every $x, y \in A$,

$$\tilde{d}_2(\eta(x),\eta(y)) = d_2(\psi \circ \eta(x), \psi \circ \eta(y)) =$$
$$= d_2(\psi \circ \eta \circ \phi^{-1}(\phi(x)), \psi \circ \eta \circ \phi^{-1}(\phi(y))) \leq$$
$$\leq d_1(\phi(x),\phi(y)) =$$
$$= \tilde{d}_1(x,y),$$

because (d_1,d_2) is a special pair. Therefore $(\tilde{d}_1,\tilde{d}_2)$ is a special pair on A, i.e.

$$\tilde{d}_1 = \tilde{d}_2 = \tilde{d}$$

is a special distance on A. As an application of Lemma 3.2 there exists a unique function $h \in S$ such that

$$\tilde{d} = d_A^h .$$

Thus making use of Theorem 1.7 one has, for all $u,v \in Q_1$:

$$d_1(u,v) = \tilde{d}(\phi^{-1}(u),\phi^{-1}(v)) =$$
$$= h(-\log K_A(\phi^{-1}(u),\phi^{-1}(v)), \log K_A(\phi^{-1}(v),\phi^{-1}(u))) =$$
$$= h(-\log K_{Q_1}(u,v), \log K_{Q_1}(v,u)) =$$
$$= d_{Q_1}^h(u,v).$$

The same arguments used for d_2 lead to the conclusion. ∎

Now we can state the main result.

<u>Theorem 3.4</u> Let $V_1 \subset R_1$ and $V_2 \subset R_2$ be two convex sharp cones, open with respect to the finite topology of the vector spaces R_1 and R_2, both having dimensions strictly greater than 1.
Let d_1 and d_2 be two distances defined on V_1 and V_2 respectively.

If (d_1,d_2) is a special pair of distances on (V_1,V_2), then there exists one, and only one, special function $f \in S$ such that

$$d_1 = d^f_{V_1}$$

and

$$d_2 = d^f_{V_2}.$$

Proof.

Let S_1 and S_2 be two-dimensional subspaces of R_1 and R_2 (respectively) such that

$$Q_1 = V_1 \cap S_1 \neq \emptyset, \quad Q_2 = V_2 \cap S_2 \neq \emptyset.$$

Since V_1 and V_2 are open for the finite topology, Q_1 and Q_2 turn out to be open, convex, sharp cones of S_1 and S_2 of dimension two; hence both of them are isomorphic to $\mathbb{R}^+_\star \times \mathbb{R}^+_\star = A \subset \mathbb{R}^2$.

For every linear map $\beta : S_1 \longrightarrow S_2$ such that $\beta(Q_1) \subset Q_2$ we can find, with the aid of Zorn's Lemma, a linear extension $\tilde{\beta} : R_1 \longrightarrow S_2$ of β such that

$$\tilde{\beta}(V_1) \subset Q_2.$$

Therefore if $d_{1|Q_1}$ and $d_{2|Q_2}$ are the restrictions of d_1 and d_2 to Q_1 and Q_2 we get, for all $u,v \in Q_1$:

$$d_{2|Q_2}(\beta(u),\beta(v)) = d_2(\tilde{\beta}(u),\tilde{\beta}(v)) \leq$$

$$\leq d_1(u,v) = d_{1|Q_1}(u,v),$$

because (d_1,d_2) is a special pair. Hence $(d_{1|Q_1}, d_{2|Q_2})$ is a special pair on (Q_1,Q_2). Therefore Lemma 3.3 implies the existence and uniqueness of $f \in S$ such that

$$d_{1|Q_1} = d^f_{Q_1}$$

and that

$$d_{2|Q_2} = d^f_{Q_2}.$$

Moreover from Lemma 1.3 we obtain, for every $u,v \in Q_1$

$$d^f_{Q_1}(u,v) = f(-\log K_{Q_1}(u,v), \log K_{Q_1}(v,u)) =$$

$$= f(-\log K_{V_1}(u,v), \log K_{V_1}(v,u)) = d^f_{V_1}(u,v),$$

using the fact that $Sp(Q_1) = S_1$. Hence

$$d_1 = d^f_{Q_1} = d^f_{V_1} \quad \text{on} \quad Q_1$$

and

$$d_2 = d^f_{Q_2} = d^f_{V_2} \quad \text{on} \quad Q_2.$$

The assertion is proved by varying the two-dimensional subspace S_1 whilst keeping S_2 fixed, and then doing the converse. ∎

Remark.

Theorem 3.4 cannot be extended to the case in which V_1 (or V_2) has dimension one, as the following example explains.

Let $h \in S$ be a special function, $x = (x_1, x_2)$ and $y = (y_1, y_2) \in A$ and let

$$d : A \times A \longrightarrow \mathbb{R}^+$$

be defined by

$$d(x,y) = d^h_A(x,y) + \left|\arctan \frac{x_2}{x_1} - \arctan \frac{y_2}{y_1}\right|.$$

The function d is obviously a distance on A. Moreover for every $f \in \mathrm{End}(\mathbb{R}^+_\star, A)$ and all $a,b \in \mathbb{R}^+_\star$ we have

$$d(f(a),f(b)) = d^h_A(f(a),f(b)) \leq d^h_{\mathbb{R}^+_\star}(a,b) \tag{9}$$

since $f(\mathbb{R}^+_\star)$ is a half-line. On the other hand

$$d^h_{\mathbb{R}^+_\star}(g(x),g(y)) \leq d^h_A(x,y) \leq \tag{10}$$

$$\leq d^h_A(x,y) + \left|\arctan \frac{x_2}{x_1} - \arctan \frac{y_2}{y_1}\right| = d(x,y),$$

for all $x,y \in A$ and all $g \in End(A, \mathbb{R}_\star^+)$. Relations (9) and (10) imply that $(d, d^h_{\mathbb{R}_\star^+})$ is a special pair of distances on $(A, \mathbb{R}_\star^+)$. Furthermore if $\phi \in GL(A)$ is defined by

$$\phi(x_1, x_2) = (3x_1, x_2)$$

it follows that

$$\phi(1,3) = (3,3)$$

$$\phi(1,1) = (3,1)$$

and since

$$|\arctan 1 - \arctan 3| \neq |\arctan \frac{1}{3} - \arctan 1|,$$

it turns out that d is not a special distance on A. Hence

$$d \neq d_A^k$$

for all $k \in S$.

By applying Theorem 3.4 to the Kobayashi-type and Carathéodory-type distances on any open convex sharp cone in a space of dimension greater than one, we obtain that

$$\gamma_V = d_V^h$$

and that

$$\delta_V = d_V^g, \tag{11}$$

where $h,g \in S$ are the special functions defined in the examples at the end of section 2. Formula (11) has already been proved directly in Theorem 1.8.

Conclusions.

With the aid of Theorem 3.4 we can give an answer to questions a),b),c) of the Preliminaries, in the case in which the cones V, V_1, V_2 are open, convex and sharp in spaces of dimension greater than one:

The special distances on V are in a one-to-one correspondence with the functions of the set S, of which we explicitly constructed a large subset (see section 2).

The special pairs of distances on the cones (V_1, V_2) are also in a one-to-one correspondence with the elements of S: each pair consists of two special distances (on V_1 and V_2 respectively) associated to the same element of S.

Finally, given any two points u,v of a cone V and a special distance d_V^f we have

$$d_V^f(u,v) = d_{V\cap\pi(u,v)}^f(u,v) .$$

This last fact follows from the proof of Theorem 3.4.

REFERENCES

[1] T. FRANZONI, Some properties of invariant distances on convex cones; Several complex variables: Proceedings of International Conferences, Cortona, Italy, 1976-77.

[2] E. HILLE, R.S. PHILLIPS, Functional Analysis and semi-groups; Am. Math. Soc. Coll. Pub., vol. XXXI.

[3] G. GENTILI, A class of invariant distances on convex cones; Symposia Mathematica, vol. XXVI (1982), 231-243.

[4] G. GENTILI, Invariant Riemannian geometry on convex cones; Tesi di Perfezionamento, Scuola Normale Superiore, Pisa (1981).

[5] W. NOLL; J.J. SCHAFFER, Orders, gauge and distance in faceless linear cones, with examples relevant to continuum mechanics and relativity; Arch. Ration. Mech. Anal. 66 (1977), 345-377.

[6] J.J. SCHAFFER, Orders, gauge and distance in faceless linear cones; Arch. Ration. Mech. Anal., 67 (1978), 305-313.

[7] E. VESENTINI, Invariant metrics on convex cones; Ann. Sc. Norm. Super. Pisa, Cl. Sci., IV. Ser., 3 (1976), 671-696.

[8] E. VESENTINI, Variations on a theme of Carathéodory; Ann. Sc. Norm. Super. Pisa, Cl. Sci., IV. Ser., 6 (1979), 39-68.

[9] E. VESENTINI, Invariant distances and invariant differential metrics in locally convex spaces; Spectral Theory, Banach Center Publications, vol. 8, Pwn-Polish Scientific Publishers, Varsaw (1982), 493-512.

SIMON SALAMON

TOPICS IN FOUR-DIMENSIONAL RIEMANNIAN GEOMETRY

PREFACE

These notes provide an introduction to some aspects of Riemannian geometry that have been developed recently, specifically the construction of the Penrose twistor space of a 4-dimensional manifold with self-dual curvature. The early sections are devoted to necessary preparatory material, sometimes without restriction to four dimensions. This ensures that the presentation is reasonably self-contained, and enables us to lead up to the relevant ideas in a natural manner. The use of fibre bundles and other fundamental concepts from differential geometry are assumed, although certain definitions are recalled partly for reasons of notation.

In order to study a particular tensor on a manifold, an essential task is to decompose the vector space containing the tensor into irreducible components with respect to the action of an appropriate Lie group. This is especially valuable in the case of torsion and curvature tensors when the components have important geometrical significance. In higher dimensions the decompositions can be a little tedious, but in four dimensions the problem is much easier because everything reduces to the group SU(2). This amounts to using spinors in preference to vectors, and is a viewpoint that we follow from the outset. Throughout special emphasis is given to explanations in terms of simple representations of Lie groups, and this approach often allows us to work independently of coordinates.

After a description in section 5 of the curvature of an oriented Riemannian 4-manifold M and the ensuing notion of self-duality, attention is focused upon almost complex structures. The existence under suitable conditions of the twistor space, a certain complex 3-manifold Z fibring over M, is established in section 8. This is one of the principal results in the paper of Atiyah, Hitchin and Singer [AHS], but our methods are a little different. Significant generalizations for higher dimensional M are now known, to which references are made when appropriate. However there is still some advantage in studying the 4-dimensional case separately. Obvious properties of the complex manifold Z are discussed in section 9, whereas the main areas of current research are represented by sections 10 and 11 in which the geometry of M and Z is interrelated. The treatment given of this has not yet appeared elsewhere. Finally, by way of conclusion, we indicate how the theory fits into a conformally invariant setting.

1. ELEMENTARY REPRESENTATION THEORY

Let M be an oriented Riemannian n-manifold. The Riemannian structure of M is encapsulated in its principal bundle P of oriented orthonomal frames, each fibre of which is isomorphic to group $SO(n)$ of $n \times n$ real matrices X with $XX^T = 1$, $\det X = 1$. If $x \in M$, a point p of the fibre P_x is a basis of T_xM and so determines an isomorphism $p : \mathbb{R}^n \longrightarrow T_xM$. Any other point of P_x has the form $pg = p \circ g$ where $g : \mathbb{R}^n \longrightarrow \mathbb{R}^n$ is the linear transformation corresponding to an element of $SO(n)$. Many objects associated to M, for instance the Riemannian connection and its geodesics, have simple interpretations on P, and the bundle P is indispensable in solving the equivalence problem. More generally, it is true to say that the study of total spaces of bundles is of paramount importance in differential geometry. In the sequel, we shall emphasize more the role played by vector bundles which are readily constructed from the principal bundle P.

By a <u>representation</u> of a Lie group G we mean a continuous homomorphism

$$\rho : G \longrightarrow \operatorname{Aut} V$$

where V is a finite-dimensional vector space, referred to as a G-<u>module</u>. This G "acts" on V, and for simplicity one writes $\rho(g)v = gv$ so that $ev = v$ where e is the identity, and $(gh)v = g(hv)$ for $g,h \in G$. A homomorphism/isomorphism $f : V \longrightarrow V'$ of G-modules is simply a vector space homomorphism/isomorphism commuting with the respective group actions. Then any $SO(n)$-module V defines an <u>associated vector bundle</u>

$$\underline{V} = P \times_{SO(n)} V$$

consisting of equivalence classes $\{p,v\}$ with $\{p,v\} = \{pg,g^{-1}v\}$ for any $g \in G$. Isomorphic $SO(n)$-modules give rise to isomorphic associated vector bundles.

Example. If $T = \mathbb{R}^n$ is the basic $SO(n)$-module determined by matrix multiplication on column vectors, then $\underline{T} \cong TM$ is the tangent bundle of M. We can perform algebraic operations to obtain new vector bundles such as the 2-forms $\Lambda^2\underline{T}^*$, but we shall meet vector bundles arising from the Riemannian structure of M which cannot be expressed directly in terms of $\underline{T}$.

In 4-dimensional Riemannian geometry the relevant group is $SO(4)$, which one studies by exploiting the fact that the Euclidean space $\mathbb{R}^4$ can be identified with the skew-field $\mathbb{H}$ of quaternions. If $q = x_0 + x_1 i + x_2 j + x_3 k \in \mathbb{H}$ and $\bar{q} = x_0 - x_1 i - x_2 j - x_3 k$ is its conjugate, then the standard metric on $\mathbb{R}^4$ is given by

$$\langle q_1, q_2\rangle = \mathrm{Re}(q_1\bar{q}_2).$$

By definition $Sp(n)$ is the group of $n \times n$ quaternionic matrices with $A\bar{A}^T = 1$, so

$$Sp(1) = \{q \in \mathbb{H} : q\bar{q} = \|q\|^2 = 1\}$$

is the group of unit quaternions. On the other hand $SU(n)$ is the group of $n \times n$ complex matrices with $A\bar{A}^T = 1$, $\det A = 1$, and the correspondence

$$q = \alpha + j\beta \longmapsto \begin{pmatrix} \alpha & -\bar{\beta} \\ \beta & \bar{\alpha} \end{pmatrix} \tag{1.1}$$

determines an isomorphism $Sp(1) \cong SU(2)$.

Proposition 1.1 There is a commutative diagram of homomorphisms

$$\begin{array}{ccc} Sp(1) & \xrightarrow{\ i\ } & Sp(1) \times Sp(1) \\ \downarrow{\scriptstyle 2:1} & & {\scriptstyle f}\downarrow{\scriptstyle 2:1} \\ SO(3) & \hookrightarrow & SO(4)\ . \end{array}$$

<u>Proof</u>. Let $(q_1,q_2) \in Sp(1) \times Sp(1)$ and define $\phi : \mathbb{H} \longrightarrow \mathbb{H}$ by $\phi(q) = q_1 q \bar{q}_2$. Since

$$\|\phi(q)\|^2 = q_1 q \bar{q}_2 \cdot q_2 \bar{q} \bar{q}_1 = \|q\|^2,$$

$\phi \in SO(4)$. Then $(q_1,q_2) \longmapsto \phi$ determines the homomorphism f, and $\ker f$ is readily seen to consist of $(1,1)$ and $(-1,-1)$. If $q_1 = q_2$, ϕ preserves the decomposition $\mathbb{H} = \mathbb{R} \oplus \mathbb{R}^3$ of q into its real and imaginary parts, and is an element of $SO(3) \subset Aut(\mathbb{R}^3)$. So to complete the diagram take i to be the diagonal inclusion $q_1 \longmapsto (q_1,q_1)$. ■

There is a diffeomorphism $Sp(1) \approx S^3$, so $Sp(1)$ is simply-connected and proposition 1.1 gives $\pi_1(SO(3)) \cong \mathbb{Z}_2 \cong \pi_1(SO(4))$. More generally the action of $SO(n+1)$ on $\mathbb{R}^{n+1}$ induces a transitive action on the sphere S^n with isotropy subgroup $SO(n)$, so there is a homogeneous description $S^n \cong {SO(n+1)}/{SO(n)}$. The exact homotopy sequence of the corresponding fibration then gives $\pi_1(SO(n)) \cong \pi_1(SO(n+1))$, $n \geq 3$. Therefore $\pi_1(SO(n)) \cong \mathbb{Z}_2$ for all $n \geq 3$, and there exists a simply-connected covering group called $Spin(n)$. We have shown

<u>Corollary 1.2</u> $Spin(3) \cong Sp(1)$, $Spin(4) \cong Sp(1) \times Sp(1)$.

This result reduces many aspects of 4-dimensional Riemannian geometry to the representation theory of the group $Sp(1)$ which we study next. This is a satisfactory state of affairs because the representation theory of compact Lie groups is best begun with the example $Sp(1)$. Left multiplication by unit quaternions determines a representation $Sp(1) \longrightarrow Aut_{\mathbb{H}} \mathbb{H}$, provided we regard $\mathbb{H}$ as a right $\mathbb{H}$-module. However it is less confusing to work over $\mathbb{C}$, so putting $V = \mathbb{C}^2$ we consider instead the underlying complex representation

$$(1.2) \qquad \rho : Sp(1) \longrightarrow Aut_{\mathbb{C}} V.$$

From (1.1), this is also the representation given by matrix multiplication of $SU(2)$ on column vectors; if $g = \begin{pmatrix} \alpha & -\bar{\beta} \\ \beta & \bar{\alpha} \end{pmatrix} \in SU(2)$, $v_1 = \begin{pmatrix} 1 \\ 0 \end{pmatrix}$,

$v_2 = \binom{0}{1}$, then $gv_1 = \alpha v_1 + \beta v_2$ and $gv_2 = -\bar{\beta} v_1 + \bar{\alpha} v_2$.

How can one find other Sp(1)-modules? Starting from V one can take tensor products, e.g.

$$V \otimes_{\mathbb{C}} V = S^2 V \oplus \Lambda^2 V$$

in which S^2V is spanned by $v_1 \otimes v_1$, $v_2 \otimes v_2$, $v_1 \otimes v_2 + v_2 \otimes v_1$, and $\Lambda^2 V$ by $\varepsilon = v_1 \wedge v_2 = v_1 \otimes v_2 - v_2 \otimes v_1$. Now Sp(1) acts by $g(u \otimes v) = gu \otimes gv$ and, using $\det_{\mathbb{C}}(g) = 1$, it follows that $g\varepsilon = \varepsilon$ for all $g \in Sp(1)$, and $\Lambda^2 V$ is a trivial Sp(1)-module. There is also the dual $V^{\star} = \mathrm{Hom}_{\mathbb{C}}(V, \mathbb{C})$ with group action $(g\phi)(v) = \phi(g^{-1}v)$, $\phi \in V^{\star}$. But $\phi \longmapsto \varepsilon(\phi) = \phi(v_1)v_2 - \phi(v_2)v_1$ is an isomorphism $V^{\star} \cong V$ of Sp(1)-modules. Finally there is always the adjoint representation of a Lie group on its Lie algebra:

$$\mathrm{Ad} : Sp(1) \longrightarrow \mathrm{Aut}_{\mathbb{R}}(sp(1)).$$

Differentiating (1.2) gives

$$sp(1) \longrightarrow \mathrm{End}_{\mathbb{C}} V \cong V^{\star} \otimes V \cong V \otimes V$$

and an isomorphism $sp(1) \cong S^2 V$ of Sp(1)-modules or more correctly

$$sp(1) \otimes_{\mathbb{R}} \mathbb{C} \cong S^2 V.$$

To clarify the distinction between real and complex spaces we need the following notion [A]. A structure map for a representation $G \longrightarrow \mathrm{Aut}_{\mathbb{C}} V$ of a Lie group G on a complex vector space V is an antilinear mapping $j : V \longrightarrow V$ commuting with the action of G such that (a) $j^2 = +1$. Then $W = \{v \in V : jv = v\}$ is a real vector space and $V \cong W \otimes_{\mathbb{R}} \mathbb{C}$ is its complexification. Thus the representation is real, and j corresponds to complex conjugation.
Or (b) $j^2 = -1$. In this case we may interpret j as a quaternion to make V into a right $\mathbb{H}$-space and the representation is quaternionic.

From its definition, the basic representation (1.2) of Sp(1) is quaternionic; its structure map j is the antilinear extension of $j(v_1) = v_2$, $j(v_2) = -v_1$.

Let S^rV denote the subspace of $\otimes^r V$ consisting of totally symmetric tensors, which may be identified with homogeneous polynomials of degree r in 2-variables. Thus $\dim_{\mathbb{C}}(S^rV) = r + 1$; e.g. S^3V has a basis

$$\{v_1 \otimes v_1 \otimes v_1,\ v_2 \otimes v_2 \otimes v_2,\ v_1 \otimes v_1 \otimes v_2 + v_1 \otimes v_2 \otimes v_1 + v_2 \otimes v_1 \otimes v_1,$$
$$v_2 \otimes v_2 \otimes v_1 + v_2 \otimes v_1 \otimes v_2 + v_1 \otimes v_2 \otimes v_2\}$$

corresponding to the polynomials x^3, y^3, x^2y, xy^2. The subspace S^rV is obviously invariant under the action of $Sp(1)$ on $\otimes^r V$, i.e. S^rV is a submodule of $\otimes^r V$ and defines a representation of $Sp(1)$. This has a structure map $\otimes^r j$ with square $(-1)^r$, so is real or quaternionic according as r is even or odd. Now S^rV is itself irreducible, i.e. contains no non-trivial $Sp(1)$-submodule. In fact

<u>Theorem 1.3</u> The set of irreducible complex $Sp(1)$-modules (up to isomorphism) is $\{S^rV : r \geq 0\}$, and

$$S^pV \otimes S^qV \cong \bigoplus_{r=0}^{\min(p,q)} S^{p+q-2r}V. \tag{1.3}$$

<u>Idea of proof</u>. To show that S^rV is irreducible suppose that $S^rV = W \oplus W'$ where W, W' are invariant by $Sp(1)$. The maximal toral subgroup $U(1)$ consisting of matrices $\begin{pmatrix} e^{it} & 0 \\ 0 & e^{-it} \end{pmatrix}$ in $SU(2) \cong Sp(1)$ leaves invariant the 1-dimensional span of $v_1^r = v_1 \otimes \ldots \otimes v_1$, so without loss of generality $v_1^r \in W$. Hence $(\alpha v_1 + \beta v_2)^r \in W$ whenever $|\alpha|^2 + |\beta|^2 = 1$. Using the Van der Monde determinant, one can choose pairs (α_i, β_i), $i = 1, \ldots, r + 1$, so that the elements $(\alpha_i v_1 + \beta_i v_2)^r$ are linearly independent. Thus $W = S^rV$.

Any representation $\rho : G \longrightarrow \mathrm{Aut}_{\mathbb{C}} V$ of a compact Lie group on a complex n-dimensional vector space is unitary, i.e. V has a basis with respect to which $\rho(G) \subset U(n)$. This follows [A] from the existence of integration on G, and implies that any G-module can be expressed as a direct sum of irreducible ones. We will illustate (1.3) for $p = 2$, $q = 1$; the general case is similar. There exist homomorphisms

$$S^2V \otimes V \xrightarrow{\text{symmetrization}} S^3V, \quad S^2V \otimes V \longrightarrow (\underbrace{V \otimes V) \otimes V}) \xrightarrow[\text{via } \varepsilon]{\text{contraction}} V.$$

Schur's lemma says that any homomorphism f between irreducible G-modules is either zero or an isomorphism (because ker f, im f are submodules!). Consequently $S^2V \otimes V$ contains submodules isomorphic to V, S^3V, and counting dimensions,

$$S^2V \otimes V \cong V \oplus S^3V.$$

The fact that the S^rV exhaust the irreducible complex Sp(1)-modules now follows from (1.3) and the Peter-Weyl theorem. A practical version of the latter [Z] states that any irreducible complex G-module for G compact can be realized as a submodule of the tensor product

$$(\otimes^p V) \otimes (\otimes^q V^*)$$

where V is any faithful (i.e. $\ker \rho = 1$) G-module. ∎

2. REPRESENTATIONS OF SO(4)

Any SO(4)-module W many be regarded as an $Sp(1) \times Sp(1)$-module by means of the homomorphism $\rho \circ f$ in the diagram

$$\begin{array}{ccc} Sp(1) \times Sp(1) & & \\ f \big\downarrow & & \\ SO(4) & \xrightarrow{\rho} & \text{Aut } W. \end{array}$$

If G_1, G_2 are compact Lie groups, then the irreducible complex $G_1 \times G_2$-modules are precisely those of the form $V_1 \otimes V_2$, where V_i is an irreducible complex G_i-module. For the proof of this statement see [A]. Using the labels +, - to distinguish betweeen the two factors of $Sp(1) \times Sp(1)$, let $V_\pm (\cong \mathbb{C}^2$ with structure map $j_\pm)$ be the corresponding $Sp(1)$-modules. Using theorem 1.3 it follows that any irreducible complex $Sp(1) \times Sp(1)$-module has the form

$$S^{p,q} = S^p V_+ \otimes S^q V_-, \qquad p,q \geq 0.$$

<u>Theorem 2.1</u> The irreducible complex SO(4)-modules are those $S^{p,q}$ with $p + q$ even, and are real representations.

<u>Proof</u>. If $1 \in SO(4)$ denotes the identity, $f^{-1}(1) = \{(1,1),(-1,-1)\}$. Since $(-1,-1)$ acts on $S^{p,q}$ as $(-1)^{p+q}$, the homomorphism $Sp(1) \times Sp(1) \longrightarrow \text{Aut}(S^{p,q})$ factors through f to give a representation of SO(4) iff $p + q$ is even. In general $S^{p,q}$ has a structure map $(\otimes^p j_+) \otimes (\otimes^q j_-)$ with square $(-1)^{p+q}$, so $S^{p,q}$ is the complexification of a real vector space for $p + q$ even. The theorem now follows from the preceding remarks. Note also that the only irreducible <u>real</u> SO(4)-modules are the $S^{p,q}$. ∎

Since $\dim_{\mathbb{C}}(S^{p,q}) = (p + 1)(q + 1)$, the basic SO(4)-module T

associated to the tangent bundle of an oriented Riemannian 4-manifold M must be $S^{1,1}$, i.e.

$$T \cong V_+ \otimes V_- .$$

In formulae like this we adopt the convention that all vector spaces, tensor products and the like are over the complex field $\mathbb{C}$. In other words a vector space which (like T) was introduced as real is automatically complexified. Similarly a quaternionic space (like V_+ if defined by $Sp(1) \longrightarrow Aut_{\mathbb{H}} \mathbb{H})$ is replaced by its underlying complex space. The real and quaternionic features are the detected by the presence of structure maps.

The Riemannian metric g determines an isomorphism $T \cong T^*$ of SO(4)-modules by the classical process of raising or lowering indices. Thus

$$\operatorname{End} T \cong T^* \otimes T^* \cong \Lambda^2 T^* \oplus \{g\} \oplus S_0^2 T^*,$$

where $S_0^2 T^*$ is the space of symmetric traceless tensors. Now

$$\Lambda^2(V_+ \otimes V_-) \cong S^2V_+ \otimes \Lambda^2 V_- \oplus \Lambda^2 V_+ \otimes S^2 V_-,$$

giving

$$(2.1) \qquad \Lambda^2 T^* \cong S^2 V_+ \oplus S^2 V_- .$$

Since $S^2V_\pm$ is the adjoint representation of $Sp(1)$, $\Lambda^2 T^*$ is the adjoint representation of $SO(4)$ and (2.1) is merely the Lie algebra splitting $so(4) \cong sp(1) \oplus sp(1)$. In general the adjoint representation of $SO(n)$ is isomorphic to $\Lambda^2 T$, and that of $Sp(n)$ is isomorphic to S^2U where $U \cong \mathbb{C}^{2n}$ is the basic $Sp(n)$-module corresponding to the inclusion $Sp(n) \subset U(2n)$.

The decomposition (2.1) may also be detected directly by means of the $\star$-operator. The latter is a homomorphism

$$\star : \Lambda^r T^* \longrightarrow \Lambda^{n-r} T^*$$

between forms on an oriented Riemannian n-manifold determined by the formula

$$\sigma \wedge \star\tau = g(\sigma,\tau)\nu, \quad \sigma,\tau \in \Lambda^r T^\star.$$

Here g is the induced metric on $\Lambda^r T^\star$, and ν the canonical n-form: if $\{e^1,\dots,e^n\}$ is any oriented orthonormal basis of $T^\star$ then by definition $\langle e^1\wedge\dots\wedge e^r, e^1\wedge\dots\wedge e^r\rangle = 1$ and $\nu = e^1\wedge\dots\wedge e^n$. It follows that

$$\star(e^1\wedge\dots\wedge e^r) = e^{r+1}\wedge\dots\wedge e^n,$$

and to evaluate $\star$ on other simple vectors, just renumber the basis. Thus $\star^2 = (-1)^{r(n-r)} = (-1)^{r(n+1)}$; putting $n = 4$ and $r = 2$ gives $\star^2 = 1$ and

$$\Lambda^2 T^\star = \Lambda^2_+ \oplus \Lambda^2_-$$

where $\Lambda^2_\pm$ are the ±1-eigenspaces of $\star$. By its definition, $\star$ commutes with the action of $SO(4)$, so we may assume that $\Lambda^2_+ \cong S^2V_+$, $\Lambda^2_- \cong S^2V$. For any oriented orthonormal basis $\{e^1,\dots,e^4\}$ of $T^\star$, put

$$\begin{aligned}
\phi^1 &= e^1\wedge e^2 + e^3\wedge e^4 & \psi^1 &= e^1\wedge e^2 - e^3\wedge e^4\\
\phi^2 &= e^1\wedge e^3 + e^4\wedge e^2 & \psi^2 &= e^1\wedge e^3 - e^4\wedge e^2 \qquad (2.2)\\
\phi^3 &= e^1\wedge e^4 + e^2\wedge e^3 & \psi^3 &= e^1\wedge e^4 - e^2\wedge e^3.
\end{aligned}$$

Then $\{\phi^i\}$, $\{\psi^i\}$ are oriented orthonormal (up to a constant) bases of Λ^2_+, Λ^2_- respectively, and the correspondence $\{e^i\} \longmapsto (\{\phi^i\},\{\psi^i\})$ gives a double covering $SO(4) \longrightarrow SO(3)\times SO(3)$.

We also have

$$S^2(V_+\otimes V_-) \cong S^2V_+\otimes S^2V_- \oplus \Lambda^2V_+\otimes\Lambda^2V_- \cong S^2V_+\otimes S^2V_- \oplus \mathbb{R},$$

giving

Proposition 2.2 There is an isomorphism $S^2_0T^\star \cong \Lambda^2_+\otimes\Lambda^2_-$.

Explicitly this isomorphism is induced from the diagram

$$\begin{array}{ccc} \Lambda^2_+ \otimes \Lambda^2_- & \overset{\cong}{\dashrightarrow} & S^2_0 T^* \\ \downarrow & & \downarrow \\ (T^* \otimes T^*) \otimes (T^* \otimes T^*) & \xrightarrow{r} & T^* \otimes T^*, \end{array}$$

where r is the contraction given in index notation by
$a_{ijkl} \longmapsto a_{ijkl} g^{jk}$.

We are now in a position to apply our knowledge of representation theory to geometry. The simplest compact 4-manifold is the sphere S^4. Just as $S^1 \approx \mathbb{R}P^1$, $S^2 \approx \mathbb{C}P^1$, so S^4 is diffeomorphic to the quaternionic projective line $\mathbb{H}P^1$. The latter may be defined as the quotient of $\mathbb{H}^2 \backslash 0$ by the group $\mathbb{H}^*$ of non-zero quaternions acting by right multiplication, with a point $x \in \mathbb{H}P^1$ corresponding to an equivalence class $[q_0, q_1] = \{(q_0 a, q_1 a) : a \in \mathbb{H}^*\}$. The mapping $x \longmapsto q_1 q_0^{-1} \in \mathbb{H}$ gives $\mathbb{H}P^1 \approx \mathbb{R}^4 \cup \{\infty\} \approx S^4$.

There are homogeneous coset space descriptions

$$S^4 = SO(5)/SO(4), \quad \mathbb{H}P^1 = Sp(2)/Sp(1) \times Sp(1).$$

For example to see the second, observe that matrix multiplication by $Sp(2)$ on quaternionic column vectors

$$\begin{pmatrix} q_0 \\ q_1 \end{pmatrix} \longmapsto \begin{pmatrix} aq_0 + bq_1 \\ cq_0 + dq_1 \end{pmatrix}, \quad \begin{pmatrix} a & b \\ c & d \end{pmatrix} \in Sp(2) \tag{2.3}$$

induces a transitive action on $\mathbb{H}P^1$. The point $[1,0]$ is fixed iff $c = 0$, forcing $b = 0$ and $a\bar{a} = d\bar{d} = 1$, so the isotropy subgroup is $Sp(1) \times Sp(1)$.

<u>Proposition 2.3</u> $Spin(5) \cong Sp(2)$ and there is a commutative diagram

$$\begin{array}{ccc} Sp(1) \times Sp(1) & \overset{i}{\hookrightarrow} & Sp(2) \\ f \downarrow & & \downarrow f' \\ SO(4) & \hookrightarrow & SO(5). \end{array}$$

Proof. To construct the double covering f', consider first the action (2.3) of $Sp(2)$ on $\mathbb{H}^2$ which defines a complex space $U \cong \mathbb{C}^4$ with a j satisfying $j^2 = -1$. Now U has a unitary basis of the form $\{u^1, u^2 = ju^1, u^3, u^4 = ju^3\}$, and one checks explicitly that

$$\omega = u^1 \wedge u^2 + u^3 \wedge u^4 \in \Lambda^2 U$$

is invariant by $Sp(2)$. ($Sp(n)$ preserves a skew 2-tensor whereas $O(n)$ preserves a symmetric one). We have already remarked that any representation $\rho : G \longrightarrow \mathrm{Aut}_{\mathbb{C}} V$ of a compact group on a complex space is necessarily unitary. If in addition V admits a structure map j, then there exists a unitary basis of V compatible with j so that

$$\rho(G) \subset O(n) = GL(n,\mathbb{R}) \cap U(n) \qquad \text{if } j^2 = +1$$

$$\rho(G) \subset Sp(\tfrac{n}{2}) = GL(\tfrac{n}{2},\mathbb{H}) \cap U(n) \qquad \text{if } j^2 = -1.$$

The space $\Lambda^2 U$ admits the structure map $j \otimes j$ with square $+1$, so there is an orthogonal sum $\Lambda^2 U = \{\omega\} \oplus \Lambda_0^2 U$ and a homomorphism

$$f' : Sp(2) \longrightarrow SO(5) \subset \mathrm{Aut}_{\mathbb{R}}(\Lambda_0^2 U).$$

The image is SO because $Sp(2)$ is connected, and $\ker f'$ (necessarily discrete, so central) must be $\{\pm 1\}$. The diagonal inclusion i corresponds to the decomposition $U = V_+ \oplus V_-$ in terms of the basic $Sp(1)$-spaces. Hence

$$\Lambda^2 U \cong \Lambda^2 V_+ \oplus (V_+ \otimes V_-) \oplus \Lambda^2 V_-.$$

and $f' \circ i$ is determined by the formula $\Lambda_0^2 U \cong (V_+ \otimes V_-) \oplus \mathbb{R}$. But the action of the sphere's isotropy subgroup $SO(4)$ on the basic $SO(5)$-module is also given by $(V_+ \otimes V_-) \oplus \mathbb{R}$ corresponding to

$$\mathbb{R}^5 = (\text{tangent space}) \oplus (\text{normal space}).$$

This shows that $f' \circ i$ factors through f as required. ∎

Let G be a closed subgroup of a Lie group K so that $M = K/G$ is a homogeneous coset space. The action of G on the tangent space

T_oM at the identity coset is called the <u>linear isotropy representation</u>, and if G is compact and connected will have the form

$$\mu : G \longrightarrow SO(n)$$

relative to some basis of T_oM. Thus T_oM admits a G-invariant metric which extends to a Riemannian metric on M by decreeing that K acts as isometries. If μ is irreducible, this metric is unique up to a constant scale factor. Proposition 2.3 then implies that S^4 and $\mathbb{H}P^1$ are isomorphic as Riemannian homogeneous spaces.

In the present work we deal exclusively with positive definite Riemannian metrics, although we indicate briefly here how the other signatures can be handled. Consider the connected groups $SO(4)$, $SO_0(3,1)$, $SO_0(2,2)$ preserving a metric on $\mathbb{R}^4$ of signature $4,2,0$ respectively. The second case is that of the Lorentzian metric on Minkowski space. In the Euclidean case the $2:1$ homomorphism $f : SU(2) \times SU(2) \longrightarrow SO(4)$ was crucial. By first complexifying and then taking other real forms, one also obtains double coverings

$$SL(2,\mathbb{C}) \times SL(2,\mathbb{C}) \longrightarrow SO(4,\mathbb{C})$$

$$SL(2,\mathbb{C}) \longrightarrow SO_0(3,1)$$

$$SU(1,1) \times SU(1,1) \longrightarrow SO_0(2,2)$$

The complexified tangent space of a Lorentzian manifold then has the form

$$T \cong V \otimes_{\mathbb{C}} \bar{V}$$

where V is the basic representation of $SL(2,\mathbb{C})$ on $\mathbb{C}^2$. For example a null vector is the product of a spinor $v \in V$ and its complex conjugate $\bar{v} \in \bar{V}$. As for the group $SU(1,1)$, this acts as automorphisms on the unit disc in $\mathbb{C}$, and identifying the latter with the upper half plane and the cone $x^2 + y^2 = z^2$, $z > 0$, respectively gives $SU(1,1) \cong SL(2,\mathbb{R}) \xrightarrow{2:1} SO_0(2,1)$. In particular the tangent module of a (++--) manifold is given by

$$T \cong V_+ \otimes_{\mathbb{R}} V_-$$

where $V_\pm$ now denotes the basic representation of $SL(2,\mathbb{R})$ on $\mathbb{R}^2$. Finally observe that complexifying (2.2) gives a double covering

$$SO(4,\mathbb{C}) \longrightarrow SO(3,\mathbb{C}) \times SO(3,\mathbb{C}).$$

However as a real space, Λ^2T^* is only reducible when the signature is 0 or 4. In the Lorentzian case, the complex summands S^2V, $S^2\bar{V}$ of Λ^2T^* each determine an isomorphism $SO_0(3,1) \cong SO(3,\mathbb{C})$.

3. SPIN MANIFOLDS

We have seen that the tangent bundle of a Riemannian 4-manifold can be expressed as the bundle

$$\underline{T} = P \times_{SO(4)} (V_+ \otimes V_-)$$

associated to the principal bundle of oriented orthonormal frames by means of a basic representation of $SO(4)$. It is tempting to write $\underline{T} \cong \underline{V}_+ \otimes \underline{V}_-$ as a tensor product of vector bundles, but in general this is meaningless because V_+, V_- are not $SO(4)$-modules, and there is no way of associating vector bundles with them globally. However it can be done for $S^4 \cong \mathbb{HP}^1$.

A homogeneous space K/G always has the principal G-bundle $\pi : K \longrightarrow K/G$ in which π projects $k \in K$ to its coset kG, and G acts on the total space K on the right preserving the fibres. For the Riemannian homogeneous space $\mathbb{HP}^1$ this gives

$$\begin{array}{c} \tilde{P} = Sp(2) \\ \pi \Big\downarrow \quad G = Sp(1)\times Sp(1) \\ \mathbb{HP}^1 . \end{array}$$

If $p = \begin{pmatrix} a & b \\ c & d \end{pmatrix} \in Sp(2)$, then taking $[1,0] \in \mathbb{HP}^1$ as the origin, π is given explicitly by $\pi(p) = [a,c]$. Since V_+, V_- <u>are</u> G-modules we can now define

$$(3.1) \qquad \underline{V}_+ = \tilde{P} \times_G V_+, \quad \underline{V}_- = \tilde{P} \times_G V_-$$

Geometrical interpretations of these vector bundles are provided in the proof of

<u>Proposition 3.1</u> The direct sum $\underline{V}_+ \oplus \underline{V}_-$ is isomorphic to the trivial vector bundle $\mathbb{HP}^1 \times \mathbb{H}^2$.

<u>Proof</u>. From the proof of proposition 2.3, the $Sp(2)$-module U decomposes as $V_+ \oplus V_-$ under the diagonal subgroup $Sp(1) \times Sp(1)$. Consequently $\underline{V}_+ \oplus \underline{V}_- \cong \tilde{P} \times_G U$ consists of equivalence classes $\{p,u\}$, $p \in Sp(2)$, $u \in U$, and $\{p,q\} = \{pg^{-1}, gu\}$ where g acts on u by

$$g \in Sp(1) \times Sp(1) \hookrightarrow Sp(2) \longrightarrow \mathrm{Aut}\, U.$$

But then $\phi\,\{p,u\} = \{\pi(p), pu\}$ defines an isomorphism $\underline{V}_+ \oplus \underline{V}_- \longrightarrow \mathbb{HP}^1 \times U \cong \mathbb{HP}^1 \times \mathbb{H}^2$.

More explicitly, if $p = \begin{pmatrix} a & b \\ c & d \end{pmatrix} \in Sp(2)$ and $u \in V_+$, then pu may be represented by $\begin{pmatrix} a & b \\ c & d \end{pmatrix} \begin{pmatrix} u \\ 0 \end{pmatrix} = \begin{pmatrix} au \\ cu \end{pmatrix}$, and lies on the quaternionic line determined by $\pi(p) = [a,c] \in \mathbb{HP}^1$. Therefore $\phi(\underline{V}_+)$ is the tautologous line bundle whose fibre over any $x \in \mathbb{HP}^1$ is simply the quaternionic line in $\mathbb{H}^2$ determined by x. Similarly $\phi(\underline{V}_-)$ is the orthogonal complement of the tautologous line bundle relative to a standard metric on $\mathbb{H}^2$. ∎

The homomorphisms $f : Sp(1) \times Sp(1) \longrightarrow SO(4)$, $f' : Sp(2) \longrightarrow SO(5)$ of proposition 2.3 induce a morphism of principal bundles

$$(3.2) \qquad \begin{array}{ccc} \tilde{P} & \xrightarrow{f'} & P \\ {\scriptstyle Sp(1)\times Sp(1)} \searrow & & \swarrow {\scriptstyle SO(4)} \\ & \mathbb{HP}^1 & \end{array} \qquad \begin{array}{l} f'(pg) = f'(p)f(g), \\ p \in \tilde{P},\ g \in Sp(1) \times Sp(1) \end{array}$$

in which P is the principal bundle corresponding to the homogeneous description $SO(5)/SO(4)$, and coincides with the bundle of oriented orthonormal frames. Given an $SO(4)$-module V, we now have two naturally associated vector bundles, namely $\tilde{P} \times_{Sp(1) \times Sp(1)} V$ and $P \times_{SO(4)} V$. But (3.2) induces an isomorphism between them, so there is no ambiguity. For example if $V = V_+ \otimes V_-$ the former is isomorphic to the tensor product $\underline{V}_+ \otimes \underline{V}_-$, whereas the latter is the tangent bundle $\underline{T}$. Hence for $\mathbb{H}P^1$,

<u>Proposition 3.2</u> $\underline{T} \cong \underline{V}_+ \otimes \underline{V}_-$.

A worthwhile exercise is to prove this directly using the realization of $\underline{V}_+, \underline{V}_-$ as subbundles of $\mathbb{H}P^1 \times \mathbb{H}^2$.

The morphism (3.2) is an example of a more general construction. Let P be any principal G-bundle on a manifold M, and $\mu : G \longrightarrow H$ a group homomorphism. Then there is an induced principal H-bundle $P_\mu = P \times_G H$ consisting of equivalence classes $\{p,h\} = \{pg, \mu(g^{-1})h\}$, and a morphism $\mu' : P \longrightarrow P_\mu$ given by $\mu'(p) = \{p,e\}$ where $e \in H$ is the identity:

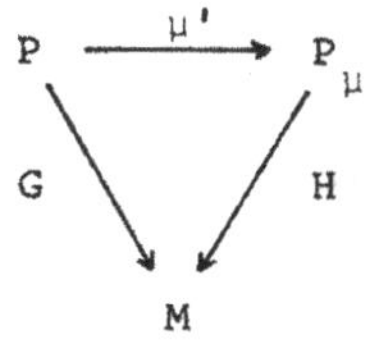

$$\mu'(pg) = \mu'(p)\mu(g),$$

$$p \in P, \ g \in G$$

Then if $M = K/G$ is Riemannian homogeneous with corresponding linear isotropy representation $\mu : G \longrightarrow SO(n)$, the induced bundle K_μ may be identified with principal SO(n)-bundle of oriented orthonormal frames, whereas the image $\mu'(K)$ consists of frames "adapted" to the homogeneous structure.

<u>Definition</u>. Let M be an oriented Riemannian n-manifold with principal bundle P of oriented orthonormal frames, and let f denote the $2:1$ homomorphism $\mathrm{Spin}(n) \longrightarrow SO(n)$. Then M is <u>Spin</u> iff there exists a principal Spin(n)-bundle $\tilde{P}$ with $\tilde{P}_f \cong P$.

If M is Spin the mapping $f' : \tilde{P} \longrightarrow P$ is a double covering, so one often says "P lifts to Spin", although the Spin structure $\tilde{P}$ may not be unique. The advantage of Spin manifolds is that using $\tilde{P}$ one can define vector bundles associated to any Spin(n)-module. Specifically in 4 dimensions we can define $\underline{V}_+, \underline{V}_-$ as in (3.1), and Proposition 3.2 will be valid. We have seen that S^4 is Spin; on the other hand

Proposition 3.3 The complex projective plane $\mathbb{C}P^2$ is <u>not</u> Spin.

<u>Proof.</u> $\mathbb{C}P^2 = \mathbb{C}^3 \backslash 0 /_{\mathbb{C}^\star}$ consists of triples $[\lambda_0, \lambda_1, \lambda_2] = [\lambda_0 a, \lambda_1 a, \lambda_2 a]$, $a \in \mathbb{C}^\star$. Matrix multiplication by $SU(3)$ on column vectors induces a transitive action on $\mathbb{C}P^2$ with isotropy subgroup at $o = [1,0,0]$ consisting of matrices

$$A = \begin{pmatrix} b & 0 & 0 \\ 0 & & B \\ 0 & & \end{pmatrix}, \quad B \in U(2), \quad b = (\det B)^{-1} \in U(1),$$

so $\mathbb{C}P^2 \cong SU(3) /_{S(U(2) \times U(1))}$. Now $A \longmapsto B$ gives $S(U(2) \times U(1)) \cong U(2)$, but A acts on $T_o \mathbb{C}P^2 \cong \mathbb{C}^2$ as $b^{-1}B$. Thus the linear isotropy representation $\mu : U(2) \longrightarrow SO(4)$ has kernel $\{ \begin{pmatrix} \omega & 0 \\ 0 & \omega \end{pmatrix} : \omega^3 = 1 \} \cong \mathbb{Z}_3$.

Those A with $\det B = 1$ fix $\begin{pmatrix} 1 \\ 0 \\ 0 \end{pmatrix} \in \mathbb{C}^3$, so $SU(3) /_{SU(2)} \cong S^5$ from which we deduce that $SU(3)$ is simply-connected.

Now suppose that $\mathbb{C}P^2$ has a Spin structure $\tilde{P}$. There is a morphism μ' from the principal bundle with total space $SU(3)$ to the oriented orthonormal frames which must lift to a continuous mapping ℓ:

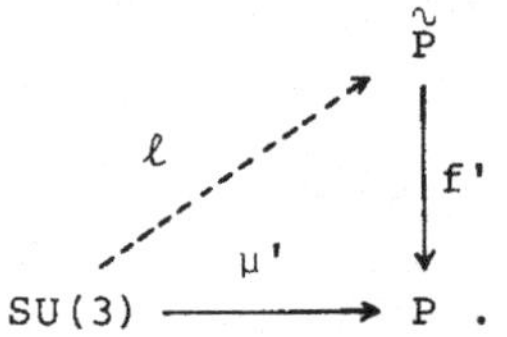

The restriction of ℓ to the fibres at o is a <u>homomorphism</u>:

$$
\begin{array}{ccc}
& & Sp(1) \times Sp(1) \\
& \ell \nearrow & \downarrow f \\
U(2) & \xrightarrow{\mu} & SO(4) \, .
\end{array}
$$

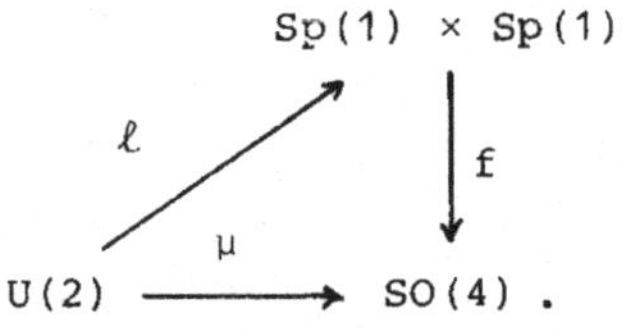

Necessarily $\ell(\ker \mu) = 1$, so there is a <u>monomorphism</u>

$U(2) /_{\ker \mu} \cong U(2) \longrightarrow Sp(1) \times Sp(1)$. This is impossible, for consider

$\sigma = \begin{pmatrix} -1 & 0 \\ 0 & 1 \end{pmatrix} \in U(2)$. Then $\sigma^2 = 1$, $\sigma \notin$ centre $U(2)$ but in $Sp(1) \times Sp(1)$ all square roots of 1 are central. ∎

To understand the topological significance of the Spin condition, one can interpret principal bundles as elements of a type of Čech cohomology group. Let M be an oriented Riemannian n-manifold with open cover $\mathcal{U} = (U_i)$, and let G be a Lie group. In this context, an n-cochain $(g_{i_0 \ldots i_n}) \in C^n$ is a collection of smooth functions

$$g_{i_0 \ldots i_n} : U_{i_0} \cap \ldots \cap U_{i_n} \longrightarrow G,$$

one for each non-empty $(n+1)$-intersection. If $(G,+)$ is abelian, one defines $d : C^n \longrightarrow C^{n+1}$ by $d(g_{i_0 \ldots i_n}) = h_{i_0 \ldots i_{n+1}}$ where

$$h_{i_0 \ldots i_{n+1}} = \sum_{k=0}^{n+1} (-1)^k g_{i_0 \ldots \hat{i}_k \ldots i_{n+1}} .$$

Then $d^2 = 0$ and one can go ahead and consider Čech cohomology

$$H^n(\mathcal{U};G) = \{\phi \in C^n : d\phi = 0\} \Big/ d(C^{n-1}) .$$

If $(G,\cdot)$ is not abelian, the best one can do is define

$$d : C^0 \longrightarrow C^1 \quad \text{by} \quad d(g_i) = (g_i g_j^{-1})$$

$$d : C^1 \longrightarrow C^2 \quad \text{by} \quad d(g_{ij}) = (g_{ij} g_{jk} g_{ik}^{-1});$$

and

$$H^1(\mathcal{U};G) = \{\phi \in C^1 : d\phi = 1\} \Big/ \sim ,$$

where $\sim$ is the equivalence relation $(g'_{ij}) \sim (g_{ij})$ iff $g'_{ij} = g_i g_{ij} g_j^{-1}$ for some $(g_i) \in C^0$. This reduces to the previous definition when G is abelian, but in general no sense can be made of H^n, $n > 1$. Regarding (g_{ij}) as transition functions, elements of $H^1(\mathcal{U};G)$ consist of isomorphism classes of principal G-bundles. $H^1(\mathcal{U};G)$ is a set with a distinguished element "1" corresponding to the trivial bundle $(g_{ij} \equiv 1)$.

To evade the process of taking direct limits, we shall suppose that $\mathcal{U}$ is sufficently nice ($U_i \subset\subset M$, each non-empty finite intersection smoothly contractible). Then the short exact sequence $1 \longrightarrow \mathbb{Z}_2 \xrightarrow{i} \mathrm{Spin}(n) \xrightarrow{f} SO(n) \longrightarrow 1$ gives

<u>Proposition 3.4</u> There is an "exact" sequence

$$\begin{array}{ccccccc} H^1(\mathcal{U};\mathbb{Z}_2) & \xrightarrow{i_\star} & H_1(\mathcal{U};\mathrm{Spin}(n)) & \xrightarrow{f_\star} & H^1(\mathcal{U};SO(n)) & \xrightarrow{d} & H^2(\mathcal{U};\mathbb{Z}_2) \\ \| & & \| & & \| & & \| \\ H^1(M,\mathbb{Z}_2) & & \text{all isomorphism} & & \text{all isomorphism} & & H^2(M;\mathbb{Z}_2) \\ \text{ordinary} & & \text{classes of prin-} & & \text{classes of prin-} & & \text{ordinary} \\ \text{cohomology} & & \text{cipal Spin(n)-} & & \text{cipal SO(n)-} & & \text{cohomology.} \\ & & \text{bundles on } M & & \text{bundles on } M & & \end{array}$$

<u>Proof.</u> The vertical equalities follow from the properties of $\mathcal{U}$ [GH,W]. The mappings $i_\star, f_\star$ are defined in an obvious manner, and coincide with the morphisms i', f' of principal bundles. If $(g_{ij}) \in C^1$ with $g_{ij}g_{jk}g_{ik}^{-1} = 1$, define $d(g_{ij}) = \tilde{g}_{ij}\tilde{g}_{jk}\tilde{g}_{ik}^{-1}$ where $\tilde{g}_{ij}$ is any lifting:

$$\begin{array}{ccc} & & \mathrm{Spin}(n) \\ & \overset{\tilde{g}_{ij}}{\nearrow} & \downarrow \\ \text{simply-connected} \quad U_i \cap U_j & \xrightarrow[g_{ij}]{} & SO(n) . \end{array}$$

One must check that d is a well-defined mapping on the cohomology level, and the resulting sequence is exact in the sense that $\mathrm{im}(i_\star) = f_\star^{-1}(1)$, $\mathrm{im}(f_\star) = d^{-1}(1)$. ■

As a corollary, M is Spin iff its principal bundle of oriented orthonormal frames P satisfies $d(P) = 1$. For example the sphere S^n, $n \geq 3$, has a unique Spin structure because $H^i(S^n,\mathbb{Z}_2) = 0$, $i = 1,2$. Because the inclusion $SO(n) \subset GL(n,\mathbb{R})$ lifts to an inclusion $\mathrm{Spin}(n) \subset \tilde{G}$ where $\tilde{G}$ is the double cover of $GL(n,\mathbb{R})$, $d(P)$ is independent of the Riemannian structure of M. It can be shown [BH] that it is the <u>2nd Stiefel-Whitney class</u> w_2 of M. Using this it follows that $\mathbb{C}P^n$ is Spin iff n is odd.

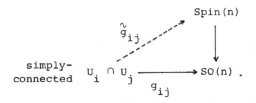

4. CONNECTIONS AND CURVATURE

In this section we review the theory of connections in principal and vector bundles. We begin by considering the homogeneous principal bundle $K \longrightarrow K/G$ where G is a closed, connected subgroup of a Lie group K, and $M = K/G$ is the space of right cosets. Let g, k denote the respective Lie algebras, and choose a complement m of g in k so that $T_eK \cong k = g \oplus m$. For any $k \in K$, the left translates $V_k = k \cdot g$, $H_k = k \cdot m$ define subspaces of T_kK such that

$$T_kK = V_k \oplus H_k.$$

The reason for choosing <u>left</u> translation is that V_k is then the vertical subspace, i.e. the tangent space to the fibre $kG = \{kg : g \in G\}$ of $K \longrightarrow M$. On the other hand each fibre kG is invariant under <u>right</u> translation by G, so it is natural to require that the horizontal subspaces H_{kg} be invariant by this action, i.e. that $H_{kg} = H_k \cdot g$ whenever $k \in K$, $g \in G$. This will be case iff

$$kg \cdot m = k \cdot m \cdot g = kg(g^{-1} m\, g),$$

i.e.

$$(4.1) \qquad (\mathrm{Ad}\, G)(m) \subseteq m$$

or equivalently $[g, m] \subseteq m$. If m can be chosen to satisfy (4.1), the homogeneous space M is said to be <u>reductive</u>. The tangent space of M at the identity coset can then be identified with m, and the linear isotropy representation with the action of $\mathrm{Ad}\, G$. M is certainly reductive in case G is either compact or semisimple.

Let $M = K/G$ be a reductive homogeneous space with $k = g \oplus m$, $[g, m] \subseteq m$. The Lie algebra structure of k determines geometrical properties of the principal bundle $K \longrightarrow M$. For example, the subalgebra condition $[\mathfrak{g},\mathfrak{g}] \subseteq \mathfrak{g}$ corresponds to the integrability of the vertical distribution V. Similarly, the $\mathfrak{g}$-component of $[\mathfrak{m},\mathfrak{m}]$ measures the non-integrability of the horizontal distribution H. To compute $[\mathfrak{m},\mathfrak{m}]$ it is convenient to introduce the $\mathfrak{k}$-valued Maurer-Cartan 1-form α on K given by

$$\alpha(X) = k^{-1}X, \quad X \in T_kK.$$

The bracket $[\mathfrak{m},\mathfrak{m}]$ is then determined by the restriction of $d\alpha$ to $\Lambda^2 H$, for

$$d\alpha(k\cdot X,k\cdot Y) = -k\cdot[X,Y], \quad X,Y \in \mathfrak{m}.$$

In general then, one expects the $\mathfrak{g}$-component of $d\alpha$ to be non-zero. On the other hand, the vanishing of the $\mathfrak{m}$-component of $d\alpha$ is precisely the condition that ensures that the homogeneous space M is <u>symmetric</u>.

When applied to an arbitrary principal bundle, the above ideas produce the concept of a <u>connection</u>. For the sake of simplicity we shall confine ourselves to the principal bundle $\pi : P \longrightarrow M$ of oriented orthonormal frames of a Riemannian n-manifold. A connection on P is then an equivariant distribution of horizontals, i.e. a distribution H satisfying

(a) $$T_pP = V_p \oplus H_p, \quad p \in P$$

(b) $$H_{pg} = H_p \cdot g, \quad p \in P,\ g \in G,$$

where V_p is the tangent space to the fibre at P. For fixed $p \in P$, the derivative of the mapping $g \longmapsto pg$ gives a natural isomorphism $so(n) \cong V_p$ which may be extended to an element $\omega \in T^*P \otimes so(n)$ by setting $H_p = \operatorname{Ker}\omega$. In addition, regarding p as a linear map $\mathbb{R}^n \longrightarrow T_{\pi(p)}M$, the composition $p^{-1} \circ \pi_*$ defines an element θ of $T^*_pP \otimes \mathbb{R}^n$ with $V_p = \ker\theta$. The $so(n) \oplus \mathbb{R}^n$-valued 1-form $\alpha = \omega + \theta$ is then the analogue of the Maurer-Cartan form above, and the restriction of $d\alpha$ to $\Lambda^2 H$ is used to define two important quantities associated to the connection. First there is the <u>curvature</u> Ω_P of P given by

$$\Omega_P(X,Y) = d\omega(hX,hY), \quad X,Y \in T_pP,$$

where h denotes "horizontal component of". To avoid the use of h, Ω_P can be computed by means of the "structure equation"

$$(4.2) \qquad \Omega_P = d\omega + [\omega,\omega].$$

Second, the <u>torsion</u> τ_P is given by

$$\tau_P(X,Y) = d\theta(hX,hY), \quad X,Y \in T_pP.$$

A well-known result states that P admits a unique connection with zero torsion, the so-called Riemannian or Levi-Civita connection. For more details, see [KN].

Next we see how the above notions carry over to a vector bundle $\underline{V} = P \times_{SO(4)} V$ corresponding to a representation $\rho : SO(n) \longrightarrow \text{Aut}\, V$. Let ω the canonical $so(n)$-valued 1-form of some connection on P. Whereas connections on principal bundles are described geometrically, those on vector bundles are usually defined in terms of a covariant derivative, i.e. a differential operator

$$\nabla : \Gamma(\underline{V}) \longrightarrow \Gamma(\underline{V} \otimes \underline{T}^{\star})$$

satisfying

$$\nabla(fv) = f\nabla v + v \otimes df,$$

where $v \in \Gamma(\underline{V})$ and f is a scalar function. As explained after theorem 2.1, one can always work with complex scalars, but then if $\underline{V}$ admits a structure map j one should require that ∇ commute with j. Take a local section of $\underline{V}$ of the form $v = \{s,\xi\}$ where $s : U \longrightarrow P$ is a section of P over some neighbourhood $U \subset M$, and $\xi \in V$ is fixed. Then there exists a unique covariant derivative ∇ characterized by

$$(4.3) \qquad \nabla\{s,\xi\} = \{s,(s^{\star}\omega)\xi\}$$

where $s^{\star}\omega$ acts on ξ via the Lie algebra homomorphism

$d\rho : so(n) \longrightarrow \text{End}\, V$. Borrowing language from physics, the local section s is called a gauge, and the $so(n)$-valued 1-form $s^\star\omega$ on U is the potential relative to s. Any other gauge has the form $s' = sg$ for some smooth function $g : U \longrightarrow SO(n)$. The fact that (4.3) provides a consistent definition of covariant differentiation follows from the formula

$$(sg)^\star\omega = \text{Ad}(g^{-1})(s^\star\omega) + g^{-1}dg$$

which implies that

$$\nabla\{s, g\xi\} = \{s, g(s^\star\omega)\xi + dg\cdot\xi\}$$

The covariant derivative ∇ has a natural extension

$$\nabla_1 : \Gamma(\underline{V} \otimes \underline{T}^\star) \longrightarrow \Gamma(\underline{V} \otimes \Lambda^2\underline{T}^\star)$$

defined in the obvious way: $\nabla_1(v \otimes \sigma) = \nabla v \wedge \sigma + v \otimes d\sigma$. The curvature of V is defined to be the composition $\Omega_V = \nabla_1\nabla$ and satisfies $\Omega_V(fv) = f\Omega_V(v)$ for f a scalar function. Thus Ω_V is a homomorphism $\underline{V} \longrightarrow \underline{V} \otimes \Lambda^2\underline{T}^\star$, or equivalently an element of $\Gamma(\text{End}\,\underline{V} \otimes \Lambda^2\underline{T}^\star)$. Using (4.2) and (4.3), it follows that Ω_V is the image of Ω_P under the homomorphism $d\rho : so(n) \longrightarrow \text{End}\,V$, and the isomorphisms $V \cong \underline{V}_x$, $H_p \cong \underline{T}_x$ corresponding to any $p \in P_x$. Relative to a local basis $\{v^i\}$ of V, the connection is expressed by

$$\nabla v^i = v^j \otimes \omega^i_j$$

for certain 1-forms ω^i_j. Then

$$\begin{aligned}\Omega_V(v^i) &= \nabla_1(v^j \otimes \omega^i_j)\\ &= \nabla v^j \wedge \omega^i_j + v^j \otimes d\omega^i_j\\ &= v^j \otimes \Omega^i_j\end{aligned}$$

where

$$\Omega^i_j = d\omega^i_j - \omega^i_k \wedge \omega^k_j. \tag{4.4}$$

The geometrical picture of a connection also carries over to vector bundles. For a covariant derivative ∇ determines a horizontal distribution H on the total space $\underline{V}$ by taking the tangent of any section $v \in \Gamma(\underline{V})$ satisfying $\nabla v|_x = 0$ to be horizontal at the point $v(x)$. If $v = \lambda_i v^i$, then

$$\nabla v = v^i \otimes (d\lambda_i + \lambda_j \omega^j_i).$$

Regarding λ_i as functions on the total space and identifying forms on M with their pullbacks, $\omega_i = d\lambda_i + \lambda_j \omega^j_i$ is a 1-form on $\underline{V}$ such that

$$\nabla v = v^i \otimes s^\star \omega_i. \tag{4.5}$$

Then $\omega_i|_H = 0$, and the ω_i play the role of the canonical form ω on P. Moreover

$$\begin{aligned} d\omega_i &= d\lambda_j \wedge \omega^j_i + \lambda_j d\omega^j_i \\ &= \omega_j \wedge \omega^j_i + \lambda_j \Omega^j_i \end{aligned} \tag{4.6}$$

Both (4.4) and (4.6) are versions of the structure equation (4.2).

Now take $V = T^\star \cong T$ and equip the cotangent bundle $\underline{T}^\star$ with the covariant derivative induced from the Riemannian connection of P. Let $\{e^i\}$ be a local oriented orthonormal basis of $\underline{T}^\star$, and put $\nabla e^i = e^j \otimes \omega^i_j$. In this case ω^i_j are simply the matrix components of $s^\star\omega$, where s is the local section of P corresponding to $\{e^i\}$; hence

$$\omega^j_i = -\,\omega^i_j. \tag{4.7}$$

By (4.3), ∇ acts on tensor products as a derivation, so

$$\nabla g = \sum_i (\nabla e^i \otimes e^i + e^i \otimes \nabla e^i) = \sum_{i,j} e^i \otimes e^j \otimes (\omega^i_j + \omega^j_i),$$

and (4.7) is equivalent to $\nabla g = 0$. Actually (4.3) implies that any tensor invariant by $SO(n)$ is covariant constant. The condition $\nabla g = 0$ is precisely the one which ensures that ∇ is induced from a connection on P.

The total space of $\underline{T}^\star$ has a tautologous 1-form whose value at $\sigma \in \underline{T}^\star$ is (the pullback of) σ itself. In coordinates this form is $\lambda_i e^i$ and is an analogue of the canonical form θ on P. Its exterior derivative is

$$\begin{aligned} d(\lambda_i e^i) &= d\lambda_i \wedge e^i + \lambda_i de^i \\ &= \omega_i \wedge e^i + \lambda_i \tau^i \end{aligned}$$

where $\tau^i = de^i + e^j \wedge \omega^i_j$ are the components of the <u>homomorphism</u>

$$\tau = d + a\nabla : \underline{T}^\star \longrightarrow \Lambda^2 \underline{T}^\star ,$$

$a : T^\star \otimes T^\star \longrightarrow \Lambda^2 T^\star$ being the anti-symmetrizing map. One can verify that using the isomorphisms $\mathbb{R}^n \cong \underline{T}^\star_x \cong H_p$ corresponding to any $p \in P_x$, τ coincides with the torsion τ_p. The existence of the Riemannian connection on P then translates into

<u>Proposition 4.1</u> The cotangent bundle $\underline{T}^\star$ admits a unique covariant derivative ∇ satisfying

$$\text{(a)} \quad \omega^j_i = -\omega^i_j , \qquad\qquad \text{(b)} \quad de^i = -e^j \wedge \omega^i_j$$

relative to any local orthonormal basis $\{e^i\}$. Differentiating (b), the curvature forms (4.4) satisfy the first Bianchi identity $e^j \wedge \Omega^i_j = 0$.

This result is readily proved by starting with any covariant derivative $\tilde{\nabla}$ satisfying (a) (constructed using a partition of unity) and seeking ∇ of the form $\omega^i_j = \tilde{\omega}^i_j + a^i_{jk} e^k$ locally. The uniqueness and existence of a^i_{jk} follows from the algebraic fact that the homomorphism $\Lambda^2 T^\star \otimes T^\star \longrightarrow T^\star \otimes \Lambda^2 T^\star$ defined by $a^i_{jk} \longmapsto a^i_{jk} - a^i_{kj}$ is an isomorphism. Since a^i_{jk} are the components of a tensor, this construction works globally. Finally we remark that the first Bianchi identity corresponds to the commutativity of the diagram

$$\begin{array}{ccccc}
\underline{T}^{\star} & \xrightarrow{\nabla} & \underline{T}^{\star} \otimes \underline{T}^{\star} & \xrightarrow{\nabla_1} & \underline{T}^{\star} \otimes \Lambda^2 \underline{T}^{\star} \\
\| & & \Big\downarrow {\scriptstyle -a} & & \Big\downarrow {\scriptstyle -a} \\
\underline{T}^{\star} & \xrightarrow{d} & \Lambda^2 \underline{T}^{\star} & \xrightarrow{d} & \Lambda^3 \underline{T}^{\star} \ .
\end{array}$$

In the next section we return to four dimensions, and examine the curvature of the Riemannian connection in more detail.

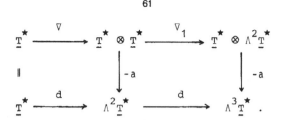

5. RIEMANNIAN CURVATURE

Let M be an oriented Riemannian 4-manifold, and let Ω denote the curvature of $\underline{T}^{\star}$ with respect to the Riemannian connection. Then Ω is a section of $\mathrm{End}\ \underline{T} \otimes \Lambda^2\underline{T}^{\star}$, but because we are only interested in the value of Ω at a given point we write simply $\Omega \in \mathrm{End}\ \underline{T} \otimes \Lambda^2\underline{T}^{\star}$. The isomorphism

$$\mathrm{End}\ T = T \otimes T^{\star} \cong T^{\star} \otimes T^{\star}$$

defined by the metric converts Ω into a tensor $R \in \underline{T}^{\star} \otimes \underline{T}^{\star} \otimes \Lambda^2\underline{T}^{\star}$ called the <u>Riemannian curvature</u>. Let $\{e^1, e^2, e^3, e^4\}$ be a local oriented orthonormal basis of 1-forms, and put $\Omega^i_j = R_{ijkl}e^k \otimes e^l$. By (4.4) and (4.7), R_{ijkl} is skew not just in k,l, but also in i,j. Thus

$$R = R_{ijkl}e^i \otimes e^j \otimes e^k \otimes e^l = \tfrac{1}{4}R_{ijkl}(e^i \wedge e^j) \otimes (e^k \wedge e^l) \in \Lambda^2\underline{T}^{\star} \otimes \Lambda^2\underline{T}^{\star}.$$

<u>Theorem 5.1</u> If $\{e^i\}$ is a local oriented orthonormal basis of $\underline{T}^{\star}$, then

$$\text{(5.1)} \qquad R = a_{ij}\ \phi^i \otimes \phi^j + b_{ij}\ \phi^i \vee \psi^j + c_{ij}\ \psi^i \otimes \psi^j \in S^2(\Lambda^2\underline{T}^{\star})$$

where ϕ^i, ψ^j are the 2-forms given by (2.2), (a_{ij}) and (c_{ij}) are symmetric with equal traces, and $\phi^i \vee \psi^j = \phi^i \otimes \psi^j + \psi^j \otimes \phi^i$.

<u>Proof</u>. Fix $x \in M$, and let Λ^r denote the $SO(4)$-module $\Lambda^r\underline{T}^{\star}$. The first Bianchi identity implies that at each point R belongs to the kernel $\mathcal{R}$ of the homomorphism

$$a : \Lambda^2 \otimes \Lambda^2 \longrightarrow \Lambda^1 \otimes \Lambda^3$$

given by $a(R_{ijkl}) = 2(R_{ijkl} + R_{iklj} + R_{iljk})$ in index notation. Thus

$$R_{ijkl} + R_{iklj} + R_{iljk} = 0$$

$$R_{ijkl} + R_{ljik} + R_{jkil} = 0$$

$$-R_{klij} - R_{ljik} - R_{iljk} = 0$$

$$-R_{klij} - R_{iklj} - R_{jkil} = 0 .$$

Adding, $2R_{ijkl} - 2R_{klij} = 0$. Thus $\mathcal{R} \subset S^2(\Lambda^2)$ and R does have the form (5.1) with $(a_{ij}),(c_{ij})$ symmetric. One checks that there is a commutative diagram

$$\begin{array}{ccc} \Lambda^2 \otimes \Lambda^2 & \xrightarrow{\ a\ } & \Lambda^1 \otimes \Lambda^3 \\ \uparrow & & \uparrow \\ S^2(\Lambda^2) & \xrightarrow{\ a'\ } & \Lambda^4 \end{array}$$

so that $\mathcal{R} = \ker a = \ker a'$. But using (2.2),

$$0 = a'(R) = a_{11} + a_{22} + a_{33} - c_{11} - c_{22} - c_{33}. \ \blacksquare$$

As an $SO(4)$-module,

$$\mathcal{R} \subset S^2(\Lambda^2_+ \oplus \Lambda^2_-) \cong S^2\Lambda^2_+ \oplus (\Lambda^2_+ \otimes \Lambda^2_-) \oplus S^2\Lambda^2_- .$$

We know that $\Lambda^2_+ \cong S^2V_+$, and from theorem 1.3, $S^2V_+ \otimes S^2V_+ \cong S^4V_+ \oplus S^2V_+ \oplus \mathbb{R}$. Counting dimensions, $S^2\Lambda^2_+ \cong S^4V_+ \oplus \mathbb{R}$; similarly for $S^2\Lambda^2_-$. The kernel of a' has codimension 1, so

<u>Corollary 5.2</u> $\mathcal{R} \cong \mathbb{R} \oplus (S^2V_+ \otimes S^2V_-) \oplus S^4V_+ \oplus S^4V_-$.

Corresponding to this irreducible decomposition (found by Singer and Thorpe [ST]), we can write

$$R = tA + B + W_+ + W_-,$$

where $t = \frac{1}{3}(a_{11} + a_{22} + a_{33})$ and

$$A = \delta_{ij}(\phi^i \otimes \phi^j + \psi^i \otimes \psi^j), \quad B = b_{ij}\phi^i \vee \phi^j$$

$$W_+ = (a_{ij} - t\delta_{ij})\phi^i \otimes \phi^j, \quad W_- = (c_{ij} - t\delta_{ij})\psi^i \otimes \psi^j.$$

Definitions. The component $W = W_+ + W_-$ is the Weyl tensor. The remaining part $tA + B$ of the curvature determines the Ricci tensor, which strictly speaking is minus the image $r(R) \in S^2\underline{T}^\star$ of R under the contraction

$$r : S^2\Lambda^2 \subset (T^\star \otimes \underbrace{T^\star) \otimes (T^\star}_{} \otimes T^\star) \longrightarrow S^2T^\star.$$

For r maps the components $(S^2V_+ \otimes S^2V_-) \oplus \mathbb{R}$ of $\mathcal{R}$ isomorphically onto $S^2T^\star$ (see proposition 2.2). In index notation, the Ricci tensor is $-R_{ijkl}g^{jk}$, and its trace $-R_{ijkl}g^{jk}g^{il} = 24t$ is the scalar curvature. In higher dimensions the Weyl and Ricci tensor are still defined, but the former is irreducible.

The Riemannian manifold M is said to be flat if $R \equiv 0$, conformally flat if $W \equiv 0$. It is well known that M is flat iff there exist local coordinates $x^1,\ldots,x^4$ such that $g = \delta_{ij}dx^i \otimes dx^j$, and conformally flat iff there exist $x^1,\ldots,x^4$ with $g = f\delta_{ij}dx^i \otimes dx^j$ for some positive function f [G]. M is Einstein if $B \equiv 0$ which means that the Ricci tensor is a multiple of the metric. These notions are all common to higher dimensions, but the special feature in 4 dimensions is the decomposition $W = W_+ + W_-$ arising from the fact that $SO(4)$ is not simple. Accordingly, M is said to be self-dual if $W_- \equiv 0$, anti-self-dual if $W_+ \equiv 0$, the distinction being a matter of orientation. The expression "conformally half-flat" is sometimes used instead of "self-dual" for reasons explained next.

There is another use of the adjective "self-dual" which unfortunately is somewhat inconsistent with the above. Suppose that F is an arbitrary vector bundle over M with a covariant derivative ∇ and curvature $\Omega_F \in \operatorname{End} F \otimes \Lambda^2\underline{T}^\star$. The latter splits as $\Omega_F = \Omega_+ + \Omega_-$ with $\Omega_+ \in \operatorname{End} F \otimes \underline{\Lambda}^2_+$, and F or ∇ is said to be self-dual if $\Omega_- \equiv 0$. The problem is that the tangent bundle $\underline{T}$ with the Riemannian connection is self-dual iff $R \in \Lambda^2\underline{T}^\star \otimes \underline{\Lambda}^2_+$, which by theorem 5.1 is the case iff

$R = W_+$, i.e. the manifold M is self-dual and Ricci flat. As another example consider the vector bundle $\underline{\Lambda}^2_+$ with its covariant derivative induced from the Riemannian connection. Since Λ^2_- is not a submodule of $\mathrm{End}\,\Lambda^2_+$, the curvature of $\underline{\Lambda}^2_+$ is determined by the component of R lying in $\underline{\Lambda}^2_+ \otimes \Lambda^2 \underline{T}^*$ which is essentially $W_+ + B + tI$. Hence $\underline{\Lambda}^2_+$ is self-dual iff M is Einstein. The same is true for the vector bundle $\underline{V}_+$ when it is defined.

Proposition 5.3 An Einstein 4-manifold M admits locally an oriented orthonormal basis $\{e^i\}$ of 1-forms so that, in the notation of (2.2),

$$R = \sum_i (a_i \phi^i \otimes \phi^i + c_i \psi^i \otimes \psi^i), \quad \Sigma\, a_i = \Sigma\, c_i .$$

Proof. The fact that M is Einstein implies that $R \in S^2\underline{\Lambda}^2_+ \oplus S^2\underline{\Lambda}^2_-$. But then there exist locally oriented orthonormal bases $\{\phi^i\}, \{\psi^i\}$ diagonalizing R; these determine $\pm\{e^i\}$ by (2.2) and the fact that $SO(4)$ double covers $SO(3) \times SO(3)$. ∎

The special basis $\{e^i\}$ was first studied by Struik [St]. With respect to it, $R_{ijkl} = 0$ whenever exactly two of the indices are equal, and the curvature tensor R is determined completely by the five quantities R_{1212}, R_{1313}, R_{1414}, R_{1234} and R_{1342}.

Examples. 1. Let $M = N_1 \times N_2$ be the Riemannian product of two surfaces. First take an oriented orthonormal basis $\{e^i\}$ of $T^*_{(x,y)}M \cong T^*_x N_1 \oplus T^*_y N_2$ such that e^1, e^2 span $T^*_x N_1$ and e^3, e^4 span $T^*_y N_2$. Then by theorem 5.1, the surfaces must have curvature

$$R_1 = k_1 (e^1 \wedge e^2) \otimes (e^1 \wedge e^2), \qquad R_2 = k_2 (e^3 \wedge e^4) \otimes (e^3 \wedge e^4)$$

where k_1, k_2 are scalar functions, called the Gaussian curvatures. Since the Riemannian covariant derivative on T^*M is the direct sum $\nabla = \nabla_1 \oplus \nabla_2$ of those on T^*N_i, M has curvature

$$R = R_1 + R_2 = \tfrac{1}{4}(k_1 + k_2)\phi^1 \otimes \phi^1 + \tfrac{1}{2}(k_1 - k_2)\phi^1 \vee \psi^1 + \tfrac{1}{4}(k_1 + k_2)\psi^1 \otimes \psi^1 .$$

If the surfaces have equal constant curvatures $k_1 = k_2$, then M is

Einstein. For example the sphere S^2 has a metric of constant positive curvature, and any compact surface of genus at least 2, expressed as a quotient of the unit disc in $\mathbb{C}$, admits a metric of constant negative curvature. On the other hand, if N_1 and N_2 have opposite constant curvatures $k_1 = -k_2$ then M is conformally flat with zero scalar curvature.

2. Let $M = K/G$ be a symmetric space with G compact, connected. This means there is a decomposition

$$\mathfrak{k} = \mathfrak{g} \oplus \mathfrak{m}; \qquad [\mathfrak{g},\mathfrak{m}] \subseteq \mathfrak{m}, \qquad [\mathfrak{m},\mathfrak{m}] \subseteq \mathfrak{g}.$$

Choose an $\mathrm{Ad}G$-invariant metric on $\mathfrak{m}$ so that the linear isotropy representation takes the form $\mu : G \longrightarrow SO(n)$; this gives M the structure of a <u>Riemannian symmetric space</u>. The connection on K determined by left-translating $\mathfrak{m}$ induces one on the principal bundle $P \cong K_\mu$ of oriented orthonormal frames which will be torsion-free because $[\mathfrak{m},\mathfrak{m}]$ has no $\mathfrak{g}$-component. This connection therefore coincides with the Riemannian connection on P whose curvature is then determined by $\Omega_P(X,Y) = [X,Y]$ for $X,Y \in \mathfrak{m}$. In other words, using $\mathfrak{m} \cong T$, $d\mu : \mathfrak{g} \longrightarrow so(4) \cong \Lambda^2T^*$, the Riemannian curvature of M is the tensor associated to the G-invariant element R in $\Lambda^2\mathfrak{m}^* \otimes \mathfrak{g} \subseteq \Lambda^2T^* \otimes \Lambda^2T^*$ corresponding to the Lie bracket $[\mathfrak{m},\mathfrak{m}]$.

If $B(X,Y) = \mathrm{tr}(\mathrm{ad}X\,\mathrm{ad}Y)$ denotes the Killing form of K, the restriction of B to $\mathfrak{g}$ is negative-definite. This is because $\mathfrak{k}$ admits an $\mathrm{Ad}G$-invariant metric relative to which $\mathrm{ad}X$ will be skew-symmetric for each $X \in \mathfrak{g}$ giving $B(X,X) = \mathrm{tr}((\mathrm{ad}X)^2) < 0$. If the linear isotropy representation (i.e. the action of $\mathrm{Ad}G$ on $\mathfrak{m}$) is irreducible, the restriction of B to $\mathfrak{m}$ must be a multiple of the chosen Riemannian metric g, both being $\mathrm{Ad}G$-invariant bilinear forms. If in addition $B|_{\mathfrak{m}} \neq 0$ so that $g = -\lambda B$ for some $\lambda \neq 0$, M is said to be <u>irreducible</u>, and in this case for $X,Y \in \mathfrak{m} \cong T$,

$$R(X,Y,X,Y) = -\lambda B([[X,Y],Y],X) = \lambda B([X,Y],[X,Y]).$$

M is said to be of compact or non-compact type according as $\lambda > 0$ or $\lambda < 0$.

According to the classification theory there are (up to homothety) two irreducible 4-dimensional Riemannian symmetric spaces of compact type, namely

$$S^4 \cong \frac{SO(5)}{SO(4)} \quad \text{and} \quad \mathbb{C}P^2 \cong \frac{SU(3)}{S(U(2) \times U(1))}.$$

Since the element A is the only $SO(4)$-invariant in $\Lambda^2 T^* \otimes \Lambda^2 T^*$, the Riemannian curvature of S^4 must equal tA for some constant $t > 0$. We shall discuss the curvature of $\mathbb{C}P^2$ in section 7.

The usefulness of corollary 5.2 in proving results expressed in classical notation is illustrated by the following which asserts that an Einstein 4-manifold also satisfies a "super-Einstein" condition:

Proposition 5.4 Let R be the Riemannian curvature of an Einstein 4-manifold. Then relative to an orthonormal basis,

$$\sum_{i,j,k} R_{ijka} R_{ijkb} = \lambda \delta_{ab}$$

for some scalar function λ.

Proof. The left hand side equals the image of $R \otimes R$ under a certain homomorphism

$$\phi : S^2(S^4V_+ \oplus S^4V_- \oplus \mathbb{R}) \subset S^2\mathcal{R} \longrightarrow T^* \otimes T^* .$$

But from theorem 1.3 and a dimension count, $S^2(S^4V_\pm) \cong S^8V_\pm \oplus S^4V_\pm \oplus \mathbb{R}$. Since the only submodule of $T^* \otimes T^*$ in common is $\mathbb{R}$, Schur's lemma implies that $\phi(R \otimes R)$ is a multiple of the metric. ■

In the notation of theorem 2.1, corollary 5.2 reads

$$\mathcal{R} \cong S^{0,0} \oplus S^{2,2} \oplus S^{4,0} \oplus S^{0,4}.$$

There is an analogous decomposition of the space $\mathcal{D}$ of derivatives of curvature tensors. Let $\nabla R \in \Lambda^2\underline{T}^* \otimes \Lambda^2\underline{T}^* \otimes \underline{T}^*$ denote the covariant derivative of the Riemannian curvature of a 4-manifold with respect to the Riemannian connection. The second Bianchi identity asserts that the image of ∇R under the anti-symmetrization

$$a : \Lambda^2T^* \otimes \Lambda^2T^* \otimes T^* \longrightarrow \Lambda^2T^* \otimes \Lambda^3T^*$$

is zero. By definition $\mathcal{D} = \ker a$. Using Schur's lemma to find out whether the restriction of a to each irreducible SO(4)-submodule is zero or an isomorphism gives in fact

<u>Proposition 5.5</u> $\mathcal{D} \cong S^{1,1} \oplus S^{3,1} \oplus S^{1,3} \oplus S^{3,3} \oplus S^{5,1} \oplus S^{1,5}$.

This result can be used to derive relations between the derivatives dt, ∇B, ∇W of the components of R. For example suppose that M is an Einstein 4-manifold, so that $B \in \underline{S}^{2,2}$ vanishes. It follows from

$$S^{2,2} \otimes S^{1,1} \cong S^{1,1} \oplus S^{3,1} \oplus S^{1,3} \oplus S^{3,3}$$

that

$$\nabla R \in \underline{S}^{5,1} \oplus \underline{S}^{1,5}.$$

In particular $dt \in \underline{S}^{1,1}$ must vanish and, as is well-known, M has constant scalar curvature. Similar remarks hold in higher dimensions, although $\mathcal{D}$ has only 4 irreducible components under the action of $SO(n)$, $n \geq 5$.

6. ALMOST HERMITIAN MANIFOLDS

Despite the fact that the tangent space to a manifold is a real object, it was convenient to consider representations over the complex numbers. This was no problem, because any vector space over $\mathbb{R}$ can be readily converted into one over $\mathbb{C}$ by complexification. In the opposite direction, starting from an arbitrary vector space over $\mathbb{C}$, the only real vector space naturally associated to it is the underlying one given by restricting scalars to $\mathbb{R}$. Let V be a complex n-dimensional G-module and let T denote the underlying real 2n-dimensional vector space. Multiplication by i on V induces a real endomorphism $I \in \operatorname{End} T$ satisfying $I^2 = -1$. Such an endomorphism is called an almost complex structure, and detects the fact that T underlies a complex space. For there is a decomposition of the complexification

$$(6.1) \qquad T_c = T \otimes_{\mathbb{R}} \mathbb{C} = T^{1,0} \oplus T^{0,1},$$

where $T^{1,0}, T^{0,1}$ are the $+i$, $-i$ eigenspaces of I respectively. For example, $T^{1,0}$ is spanned by elements of the form $X - iIX$, $X \in T$. The composition of the inclusion $V \longrightarrow T \subset T_c$ with the projection $T_c \longrightarrow T^{1,0}$ is complex linear, whereas $V \longrightarrow T^{0,1}$ is antilinear; thus $T^{1,0} \cong V$, $T^{0,1} \cong \bar{V}$.

If G is compact, V must admit a unitary basis $\{v^1,\dots,v^n\}$, i.e. one for which $g \in G$ acts as a matrix $A = (a^r_s)$ with $A\bar{A}^T = 1$. With respect to the basis $\{v^1, iv^1, \dots, v^n, iv^n\}$, g acts on T as the matrix B formed by replacing each $a^r_s = x^r_s + iy^r_s$ by $\begin{pmatrix} x^r_s & y^r_s \\ -y^r_s & x^r_s \end{pmatrix}$.

Then $BB^T = 1$ and $\det B = 1$, so $A \longmapsto B$ defines a monomorphism $\rho : U(n) \hookrightarrow SO(2n)$ whose image is the subgroup of $SO(2n)$ commuting with I. In (6.1) T can now be interpreted as the basic representation of $SO(2n)$ on $\mathbb{R}^{2n}$, and $T^{1,0}$ the basic representation of $U(n)$ on $\mathbb{C}^n$ given by matrix multiplication.

A (necessarily even dimensional) manifold M is said to be almost complex if each tangent space admits an almost complex structure I, which varies smoothly. Any complex manifold has this property, for if $z_r = x_r + iy_r$ are local holomorphic coordinates, I is given by

$$I(\frac{\partial}{\partial x_r}) = \frac{\partial}{\partial y_r}, \quad I(\frac{\partial}{\partial y_r}) = -\frac{\partial}{\partial x_r}.$$

Now suppose that M admits a Riemannian metric g compatible with the almost complex structure in the sense that

$$g(IX,IY) = g(X,Y), \quad X,Y \in \underline{T}. \tag{6.2}$$

Such a g can always be found, for example by taking $g(X,Y) = h(X,Y) + h(IX,IY)$ with h an arbitrary metric. Then (6.2) ensures that each tangent space has an orthonormal frame of the form $\{X_1, IX_1, \ldots, X_n, IX_n\}$, and the set of all such frames constitutes a principal $U(n)$-bundle Q such that Q_ρ can be identified with the $SO(2n)$-bundle of orthonormal frames oriented consistently with I. A manifold equipped with a metric g and an almost complex structure I satisfying (6.2) is called <u>almost Hermitian</u>. For example $\mathbb{C}P^2$ has an almost Hermitian structure defined by the $U(2)$-bundle $Q = \mu'(SU(3))$ (see the proof of propsition 3.3).

Let M be an almost Hermitian manifold. The bundle Q can be used to construct vector bundles associated to any $U(n)$-module. Taking the dual of (6.1) and returning to our convention of complexifying real modules on sight, the cotangent bundle has the form

$$\underline{T}^\star = \underline{\Lambda}^{1,0} \oplus \underline{\Lambda}^{0,1},$$

where $\Lambda^{1,0}, \Lambda^{0,1}$ are the annihilators of $T^{0,1}, T^{1,0}$ respectively. Actually g induces isomorphisms

$$\Lambda^{1,0} \cong T^{0,1}, \quad \Lambda^{0,1} \cong T^{1,0}, \tag{6.3}$$

corresponding to the fact that any complex unitary G-module satisfies $V^\star \cong \bar{V}$. More generally the r-forms on M decompose as

$$\Lambda^r \underline{T}^\star = \bigoplus_{p+q=r} \underline{\Lambda}^{p,q},$$

with $\Lambda^{p,q} \cong \Lambda^p(\Lambda^{1,0}) \otimes \Lambda^q(\Lambda^{0,1})$. In general $\Lambda^{p,q}$ is not an irreducible U(n)-module. For example M has a "fundamental 2-form" F given by

$$F(X,Y) = g(IX,Y)$$

which defines an element of $\Lambda^{1,1}$ invariant by U(n). Next we consider the covariant derivative $\nabla F \in \Lambda^2 \underline{T}^\star \otimes \underline{T}^\star$ of F with respect to the Riemannian connection.

<u>Lemma 6.1</u> $\nabla F \in (\underline{\Lambda}^{2,0} \oplus \underline{\Lambda}^{0,2}) \otimes \underline{T}^\star$.

<u>Proof</u>. The almost Hermitian structure of M is determined by g and F, and corresponding to $I^2 = -1$ there is the compatibility relation

$$r(F \otimes F) = -g,$$

where $r : S^2(\Lambda^2 T^\star) \longrightarrow S^2 T^\star$ is the contraction used to define the Ricci tensor in section 5. Differentiating,

$$r(F \vee \nabla_X F) = 0$$

for any $X \in \underline{T}$, where of course $\nabla_X : \Gamma(\Lambda^2 \underline{T}^\star) \longrightarrow \Gamma(\Lambda^2 \underline{T}^\star)$ is the operator formed by contraction with X. But

$$r(F \vee \nabla_X F)(Y,Z) = (\nabla_X F)(Y,IZ) + (\nabla_X F)(Z,IY).$$

Replacing Y by IY,

$$(\nabla_X F)(IY,IZ) + (\nabla_X F)(Y,Z) = 0$$

which (compare (6.2)) is equivalent to $\nabla_X F \in \underline{\Lambda}^{2,0} \oplus \underline{\Lambda}^{0,2}$. ∎

The tensor ∇F is important because it measures the extent to which the Riemannian covariant derivative ∇ fails to preserve the subbundle $\Lambda^{1,0}$ of $\underline{T}^\star$. More precisely, consider the composition

$$\eta : \underline{\Lambda}^{1,0} \xrightarrow{\nabla} \underline{T}^{\star} \otimes \underline{T}^{\star} \xrightarrow{\text{projection}} \underline{\Lambda}^{0,1} \otimes \underline{T}^{\star}$$

which is a homomorphism (it's an example of a "2nd fundamental form"). Now η may be regarded as a section of $\underline{\Lambda}^{0,1} \otimes \underline{\Lambda}^{0,1} \otimes \underline{T}^{\star}$ with $\eta(\alpha,\beta,X) = g(\nabla_X \alpha,\beta)$ for $\alpha,\beta \in \underline{T}^{0,1} \cong \underline{\Lambda}^{1,0}$, but in fact

<u>Proposition 6.2</u> $\eta \in \underline{\Lambda}^{0,2} \otimes \underline{T}^{\star}$ and $\nabla F = -2i(\eta - \bar{\eta})$.

<u>Proof</u>. If $\alpha,\beta \in \Gamma(\underline{T}^{0,1})$ so that $F(\alpha,\beta) = 0 = g(\alpha,\beta)$, then

$$\begin{aligned}(\nabla_X F)\ (\alpha,\beta) &= F(\nabla_X \alpha,\beta) + F(\alpha,\nabla_X \beta)\\ &= -ig(\nabla_X \alpha,\beta) + ig(\alpha,\nabla_X \beta)\\ &= -2ig(\nabla_X \alpha,\beta)\\ &= -2i\eta(\alpha,\beta,X),\end{aligned}$$

and the second result follows from the reality of F. ■

Write

$$\eta = \eta_c + \eta_h,$$

where $\eta_c \in \underline{\Lambda}^{0,2} \otimes \underline{\Lambda}^{0,1}$, $\eta_h \in \underline{\Lambda}^{0,2} \otimes \underline{\Lambda}^{1,0} \cong \underline{\Lambda}^{1,2}$ (the subscripts stand for "complex" and "harmonic" for reasons soon to be apparent). For any $\alpha \in \Gamma(\underline{\Lambda}^{1,0})$, the (0,2)-component of α is

$$(d\alpha)^{0,2} = (-a\nabla\alpha)^{0,2} = -a\eta_c(\alpha).$$

But $\alpha \longmapsto (d\alpha)^{0,2}$ is essentially the Nijenhuis tensor associated to the almost complex structure of M. It follows that this tensor can be identified with η_c. By the Newlander-Nirenberg theorem [NN], the almost complex structure is integrablc, i.e. M is a complex manifold, iff $\eta_c \equiv 0$.

<u>Definition</u>. An almost Hermitian manifold is said to be Kähler if $\eta \equiv 0$, or equivalently if $\nabla F \equiv 0$.

An almost Hermitian manifold is therefore Kähler iff ∇ is a connection on the vector bundle $\underline{\Lambda}^{1,0}$, which means that ∇ is induced from a connection on the principal $U(n)$-bundle Q. This is certainly the case for the Riemannian symmetric space $\mathbb{C}P^n$ which is then Kähler.

The formula

$$dF = -a\nabla F = 2ia(\eta - \bar{\eta})$$

implies that η_h can be identified with the (1,2)-component of the 3-form dF. Consequently to put a Kähler structure on a complex manifold it suffices to find a closed (1,1)-form F such that the bilinear form $F(IX,Y)$ is positive definite. The vanishing of η_h is important in the theory of harmonic mappings, for a holomorphic mapping between any two almost Hermitian manifolds, both of which have $\eta_h = 0$, is necessarily harmonic [EL]. Actually in 4 dimensions $\Lambda^3 T^* \cong T^*$, $\Lambda^{2,1} \cong \Lambda^{0,1}$, $\Lambda^{3,0} = 0$, and the tensors η_c, η_h lie in irreducible $U(2)$-modules. However in dimension $2n$, $n \geq 3$, both η_c and η_h split into two further components [GrH].

Examples. 1. There are topological obstructions to the existence of an almost complex structure on an even dimensional manifold. On a compact 4-manifold, a necessary condition is that $1 - b_1 + b_2^+$ be even, where the Betti numbers b_1, b_2^+ equal the dimensions of the spaces $\{\phi \in \Gamma(\underline{T}^*) : d\phi \equiv 0 \equiv d{\star}\phi\}$, $\{\phi \in \Gamma(\underline{\Lambda}_+^2) : d\phi \equiv 0\}$ respectively. This is a consequence of the Atiyah-Singer index theorem [AS] which implies that the Todd genus must be an integer. The latter can be expressed as $\frac{1}{4}(\chi + \tau)$ where χ is the Euler characteristic, and τ the signature.

In particular the sphere S^4 cannot admit an almost complex structure; in fact S^n can only admit one for $n = 2$ or 6. On S^2 any Riemannian metric defines a Kähler structure biholomorphically equivalent to the complex projective line $\mathbb{C}P^1$. On the other hand the Riemannian homogeneous space $G_2/SU(3) \approx S^6$ has a natural almost Hermitian structure with $\eta_h = 0$ but $\eta_c \neq 0$ [Gr].

2. A compact Kähler manifold must have its even Betti numbers non-zero (because of the existence of the closed 2-form F), and its odd Betti numbers even (see e.g. [G]). Let $\mathbb{Z}$ denote the cyclic subgroup of $\mathbb{C}^\star$ generated by some λ, $|\lambda| > 1$. Then the Hopf surface $\mathbb{C}^2\setminus 0/_{\mathbb{Z}} \approx S^3 \times S^1$ is a simple example of a complex manifold which cannot admit a Kähler metric. A more interesting example is furnished by replacing S^3 by the compact 3-manifold $M = G/_\Gamma$ where G is the Heisenberg group of matrices

$$\begin{pmatrix} 1 & a & c \\ 0 & 1 & b \\ 0 & 0 & 1 \end{pmatrix}, \quad a,b,c \in \mathbb{R}$$

under multiplication, and Γ the subgroup with $a,b,c \in \mathbb{Z}$. M is homeomorphic to the $S^1 \times S^1$-bundle over S^1 formed by the identification

$$(e^{2\pi i a}, e^{2\pi i c}, 0) \sim (e^{2\pi i a}, e^{2\pi i (c + a)}, 1)$$

in $S^1 \times S^1 \times [0,1]$. The loops $0 \le a \le 1$, $0 \le b \le 1$ generate the fundamental group $\pi_1(M) \cong \mathbb{Z} \oplus \mathbb{Z}$; thus $M \times S^1$ has $b_1 = 3$ and cannot be Kähler.

An almost Hermitian structure is specified by a Riemannian metric g and a 2-form F satisfying $r(F \otimes F) = -g$ (see lemma 6.1). The forms $dc - bda$, da, db are right-invariant on G, and so pass to forms e^1, e^2, e^3 respectively on M. Take $e^4 = dt$ on $S^1 \cong \{e^{2\pi i t} : t \in [0,1]\}$ and declare $\{e^1, \ldots, e^4\}$ to be an oriented orthonormal basis of 1-forms on $M \times S^1$. Then

$$F = e^1 \wedge e^2 + e^3 \wedge e^4, \quad F' = e^1 \wedge e^4 + e^2 \wedge e^3$$

define two different almost Hermitian structures on $M \times S^1$. Because $de^1 = e^2 \wedge e^3$ and $de^i = 0$ for $i \neq 1$, the first has $dF = 0$, so $\tau_h = 0$ and $\tau_c \neq 0$. The closed 2-form F of maximal rank defines what is called a symplectic structure, and this example of a non-Kähler symplectic manifold is due to Thurston [T]. In the second case, the (1,0)-forms relative to F' are spanned by

$$\theta^1 = e^1 + ie^4, \quad \theta^2 = e^2 + ie^3$$

which satisfy

$$d\theta^1 = -\frac{1}{2}i\theta^2 \wedge \bar{\theta}^2 \in \underline{\Lambda}^{1,1}, \quad d\theta^2 = 0.$$

This makes $M \times S^1$ into a complex manifold, so $\eta_c = 0$ and $\eta_h \neq 0$. The complex structure really arises from the identification of the group G with the real hypersurface $\operatorname{Im} z_2 = |z_1|^2$ in $\mathbb{C}^2$ given by

$$2z_1 = (a + b) + i(a - b), \quad 2z_2 = 2c - ab + i(a^2 + b^2).$$

7. REPRESENTATIONS OF U(2)

Any SO(4)-module becomes a U(2)-module via the inclusion $\rho : U(2) \hookrightarrow SO(4)$, and if irreducible under SO(4), may or may not reduce under U(2). For example the tangent space T remains irreducible over $\mathbb{R}$, but splits as $T = T^{1,0} \oplus T^{0,1}$ over $\mathbb{C}$. An excellent example of the interplay between U(2) and SO(4) is provided by the space of 2-forms, which decompose as $\Lambda^2 T^* = \Lambda^2_+ \oplus \Lambda^2_-$ under SO(4). Relative to U(2) we can certainly write

$$\Lambda^2 T^* = (\Lambda^{2,0} \oplus \Lambda^{0,2}) \oplus \Lambda^{1,1},$$

where the right hand side is the direct sum of two <u>real</u> vector spaces of dimension 2,4 respectively. But since $\Lambda^{1,1}$ contains the invariant F, there is an <u>orthogonal</u> decomposition $\Lambda^{1,1} = \{F\} \oplus \Lambda^{1,1}_0$.

<u>Proposition 7.1</u> $\Lambda^2_+ = \{F\} \oplus (\Lambda^{2,0} \oplus \Lambda^{0,2})$, $\Lambda^2_- = \Lambda^{1,1}_0$.

<u>Proof.</u> The basic SO(4)-module T admits an oriented orthonormal basis of the form $\{e_1, e_2 = Ie_1, e_3, e_4 = Ie_3\}$ where I is the almost complex structure coming from the action of U(2). If $\{e^1, e^2, e^3, e^4\}$ is the dual basis of T^*, the space $\Lambda^{1,0}$ of (1,0)-forms is generated by

$$\theta^1 = e^1 + ie^2, \quad \theta^2 = e^3 + ie^4,$$

and the result follows from (2.2) and the formulae

$$F = \tfrac{1}{2}i(\theta^1 \wedge \bar\theta^1 + \theta^2 \wedge \bar\theta^2) = \phi^1$$

$$\tfrac{1}{2}i(\theta^1 \wedge \bar\theta^1 - \theta^2 \wedge \bar\theta^2) = \psi^1$$

$$\theta^1 \wedge \theta^2 = \phi^2 + i\phi^3, \qquad \theta^1 \wedge \bar\theta^2 = \psi^2 - i\psi^3. \blacksquare$$

<u>Theorem 7.2</u> A Kähler 4-manifold has Riemannian curvature

$$R = B + tC + W_-,$$

where C is a tensor representing the curvature of $\mathbb{C}P^2$, and t, B, W_- are as in corallary 5.2. Moreover $B = F \vee \psi$ for some $\psi \in \underline{\Lambda}_-^2$.

<u>Proof</u>. For a Kähler manifold the Riemannian covariant derivative preserves $\underline{\Lambda}^{1,0}$. Hence the real cotangent bundle $\underline{T}^*$ has curvature $\Omega + \bar{\Omega}$, where Ω is the composition

$$\underline{\Lambda}^{1,0} \xrightarrow{\nabla} \underline{\Lambda}^{1,0} \otimes \underline{T}^* \xrightarrow{\nabla_1} \underline{\Lambda}^{1,0} \otimes \Lambda^2\underline{T}^*.$$

Then $\Omega \in (\underline{\Lambda}^{1,0})^* \otimes \underline{\Lambda}^{1,0} \otimes \Lambda^2\underline{T}^* \cong \Lambda^{1,1} \otimes \Lambda^2\underline{T}^*$. But always $R \in S^2(\Lambda^2\underline{T}^*)$, forcing

(7.1) $$R \in S^2(\underline{\Lambda}^{1,1}).$$

Actually $\Lambda^{1,1}$ is isomorphic to the adjoint representation of $U(n)$, and (7.1) expresses the fact that the <u>holonomy group</u> of a Kähler manifold lies in $U(n)$ [KN].

In 4 dimensions, following theorem 5.1 and proposition 7.1, we may write

$$R = a_{11}\phi^1 \otimes \phi^1 + b_{1j}\phi^1 \vee \psi^j + c_{ij}\psi^i \otimes \psi^j$$

$$= t(3F \otimes F - \delta_{ij}\psi^i \otimes \psi^j) + F \vee (b_{1j}\psi^j) + (c_{ij} - t\delta_{ij})\psi^i \otimes \psi^j.$$

The tensor $C = 3F \otimes F - \delta_{ij}\psi^i \otimes \psi^j$ is clearly invariant by $U(2)$ and the discussion in example 2 of section 5 implies that the Riemannian curvature of the Kähler manifold $\mathbb{C}P^2$ is tC with t a <u>constant</u>. ∎

In the spirit of corollary 5.2, the above implies that the space of curvature tensors at each point of a Kähler 4-manifold corresponds to the module $\mathbb{R} \oplus S^2V_- \oplus S^4V_-$. It will soon be clear that these components are all irreducible even with respect to $U(2)$. The situation is illustrated diagramatically below left. The circular segments are the irreducible SO(4)-modules that make up the space $\mathcal{R}$ and determine the Ricci and Weyl tensors, whereas the shaded areas represent the U(2)-modules containing the Kähler curvature.

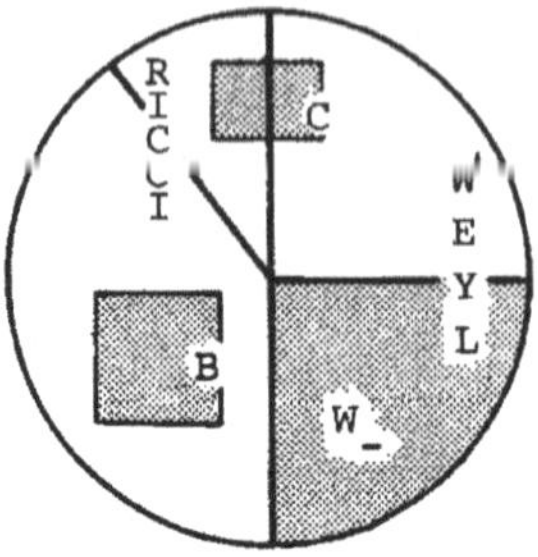

dim M = 4

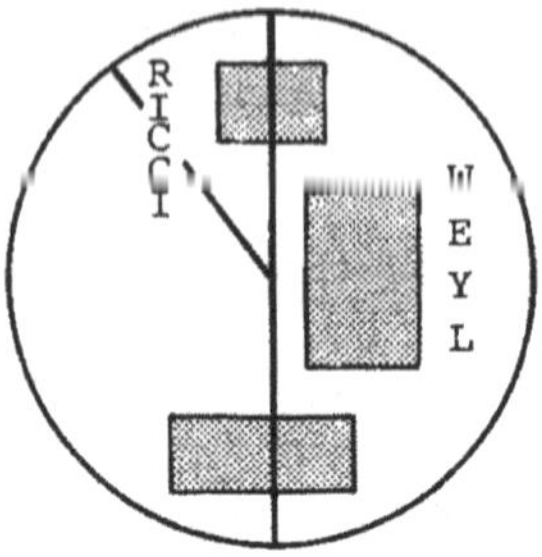

dim M = 2n, n ⩾ 3

To emphasize the special features of 4 dimensions, what happens in higher dimensions is shown alongside. The Weyl tensor no longer splits into two, so the Riemannian curvature is described by three irreducible SO(2n)-modules. Using (7.1) one can show that the curvature of a Kähler manifold is still contained in three irreducible U(n)-submodules, but the relative position of these is somewhat different.

<u>Examples</u>. 1. A K3 surface is a compact complex surface with $b_1 = 0$, $c_1 = 0$, and a simple algebraic example is provided by any quartic in $\mathbb{C}P^3$. The vanishing of the first Chern class c_1 is equivalent to the <u>canonical bundle</u> $\kappa = \underline{\Lambda}^{2,0}$ being topologically trivial. If M is a Kähler K3 surface, Yau's proof [Y] of the Calabi conjecture then implies that M admits a Kähler metric with zero Ricci tensor. From theorem 7.2, this means that $R = W_-$ and M is <u>anti</u>-self-dual with respect to our convention regarding the orientation determined by an almost complex structure. Moreover the Riemannian curvature lies in $\underline{\Lambda}^2_+ \otimes \Lambda^2\underline{T}^\star$, so the vector bundle $\underline{\Lambda}^2_-$ with its induced Riemannian connection is flat. By proposition 7.1, the same is true of the canonical bundle κ. Conversely, it can be shown [H_1] that any compact Riemannian 4-manifold which is anti-self-dual and Ricci-flat has universal covering a K3 surface, or is flat.

2. Now suppose that M is a <u>self</u>-dual Kähler manifold, so that $R = B + tC$. If the scalar curvature t is constant, then using the second Bianchi identity and the methods of proposition 5.5, it can be shown that $\nabla R = 0$. This means [KN; XI, 6] that M is locally sym-

metric, i.e. isomorphic to a neighbourhood of a Riemannian symmetric space. However if t is unknown, this conclusion still holds if M is compact [D]. An example with $B \neq 0$ is furnished by the product of two surfaces with opposite curvatures considered in section 5. This case is of particular interest because it is an example of a conformally flat Kähler manifold which is not flat, a phenomenon which cannot occur in higher dimensions.

To understand the inclusion $\rho : U(2) \hookrightarrow SO(4)$ fully, recall (proposition 1.1) that left and right multiplication by unit quaternions on $\mathbb{H}$ defines a double covering $Sp(1) \times Sp(1) \longrightarrow SO(4)$. Declare $\{1,i,j,k\}$ to be an orthonormal basis of $\mathbb{H} \cong \mathbb{R}^4$, and let I be the almost complex structure given by left multiplication by i. Then $\rho(U(2))$ is the subgroup of $SO(4)$ commuting with I. But this subgroup is double-covered by $U(1) \times Sp(1)$ where $U(1) = \{z \in \mathbb{C} : |z| = 1\}$, and there is a commutative diagram

$$\begin{array}{ccc} U(1) \times Sp(1) & \hookrightarrow & Sp(1) \times Sp(1) \\ \big\downarrow 2:1 & & \big\downarrow 2:1 \\ U(2) & \overset{\rho}{\hookrightarrow} & SO(4) \end{array}$$

Using (1.1), $z \in U(1)$ acts as the complex matrix $\begin{pmatrix} z & 0 \\ 0 & \bar{z} \end{pmatrix}$ on the basic complex $Sp(1)$-module V_+. If L denotes the eigenspace on which z acts as z, the structure map j of V_+ gives a decomposition

$$V_+ = L \oplus (jL) \cong L \oplus \bar{L}. \tag{7.2}$$

Since any $SO(4)$-module is built up from V_+ and V_-, the manner in which it breaks up under $U(2)$ is completely determined by (7.2) and the fact that V_- remains intact. For example

$$T \cong V_+ \otimes V_- = (L \otimes V_-) \oplus (jL \otimes V_-)$$

and

$$T^{1,0} \cong L \otimes V_-. \tag{7.3}$$

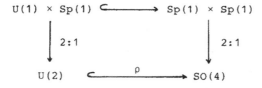

The subspace $L \subset V_+$ determines a point of the complex projective line $\mathbb{P}(V_+) = V_+ \setminus 0 / \mathbb{C}^*$. Pick any $\ell \in L$; the real element $i(\ell \vee j\ell) \in S^2V_+$, when normalized, is an element of the 2-sphere $\mathbb{S}(\Lambda^2_+) = \{\alpha \in \Lambda^2_+ : \|\alpha\| = 1\}$. This element is a (1,1)-form, so by proposition 7.1 is the fundamental 2-form F (adjusting our conventions so that $\|F\| = 1$). In fact

<u>Proposition 7.3</u> There are natural isomorphisms

$$SO(4)/U(2) \cong Sp(1)/U(1) \cong \mathbb{P}(V_+) \cong \mathbb{S}(\Lambda^2_+).$$

<u>Proof</u>. The first isomorphism comes from the diagram. Both V_+ and Λ^2_+ are $Sp(1)$-modules, and the other two isomorphisms follow by associating to the coset $gU(1) \in Sp(1)/U(1)$ the elements $gL \in \mathbb{P}(V_+)$ and $gF \in \mathbb{S}(\Lambda^2_+)$. ∎

Let M be an oriented Riemannian 4-manifold with principal bundle of oriented orthonormal frames P. A subordinate almost Hermitian structure on M is determined by a principal $U(2)$-bundle Q with $Q_\rho \cong P$. Then Q can be identified with a section of the associated homogeneous space bundle $Z = P \times_{SO(4)} SO(4)/U(2)$ by mapping $y \in Q$ to $\{\rho'(y), U(2)\}$, where $U(2)$ denotes the identity coset. The fibre Z_x parametrizes all those almost complex structures on $\underline{T}_x$ compatible with the metric and orientation, i.e. ones I for which $\underline{T}_x$ has an oriented orthonormal basis $\{X_1, IX_1, X_2, IX_2\}$.

Let $\mathbb{S}(\underline{\Lambda}^2_+)$ denote the associated bundle $P \times_{SO(4)} \mathbb{S}(\Lambda^2_+)$; it is obviously isomorphic to the bundle of unit vectors in $\underline{\Lambda}^2_+$. Similarly, let $\mathbb{P}(\underline{V}_+)$ denote $P \times_{SO(4)} \mathbb{P}(V_+)$. If M is Spin, the latter is isomorphic to the complex projective bundle $\underline{V}_+ \setminus \underline{0} / \mathbb{C}^*$, where $\underline{V}_+$ is constructed by choosing any lift of P to Spin, and $\underline{0}$ denotes the zero section. This description of $\mathbb{P}(\underline{V}_+)$ is certainly available on any sufficiently small neighbourhood U of M.

<u>Corollary 7.4</u> Each point of $Z \cong \mathbb{S}(\underline{\Lambda}^2_+) \cong \mathbb{P}(\underline{V}_+)$ determines an almost Hermitian structure on the tangent space of M below.

In higher dimensions, on an oriented Riemannian 2n-manifold M, one can still define the bundle $Z = P \times_{SO(2n)} SO(2n)/U(n)$, each fibre of which consists of all almost complex structures compatible with the metric and orientation. Associating to an almost complex structure its fundamental 2-form realizes Z as a subbundle of the 2-forms $\Lambda^2 T^* M$. For example when $n = 3$, the fibre

$$\frac{SO(6)}{U(3)} \cong \frac{SU(4)}{S(U(3) \times U(1))} \cong \mathbb{C}P^3$$

is again a projective space, and embeds in $S^{14} \subset \Lambda^2(\mathbb{R}^6)$. In general a point $z \in Z_x$ determines a decomposition

$$\Lambda^2 T_x^* M = \Lambda_z^{2,0} \oplus \Lambda_z^{1,1} \oplus \Lambda_z^{0,2},$$

in which $\Lambda_z^{2,0} \oplus \Lambda_z^{0,2}$ is naturally isomorphic to the tangent space $T_z(Z_x)$ to the fibre, and $\Lambda_z^{1,1} \cong u(n)$ is the normal space in $\Lambda^2 T_x^* M$. This generalizes the decomposition of Λ_+^2 in proposition 7.1.

The following result is contained in [AHS], and is based on ideas of Penrose, whose twistor program attempts to intepret many aspects of real 4-dimensional geometry in holomorphic terms.

Theorem 8.1 Let M be an oriented Riemannian 4-manifold which is anti-self-dual, i.e. $W_+ \equiv 0$. Then the total space of the bundle Z is a complex manifold.

Proof. Take a neighbourhood U of M over which the SO(4)-bundle P of oriented orthonormal frames is trivial. Over U, P lifts to a Spin(4)-bundle $\tilde{P}$, and the vector bundles $\underline{V}_\pm = \tilde{P} \times_{Spin(4)} V_\pm$ are defined. Let $Y = \underline{V}_+ \backslash \underline{0}$ denote the total space $\underline{V}_+$ minus its zero section. We now have projections $\pi : Z \longrightarrow M$, $\mu : Y \longrightarrow U$ and $\nu : Y \longrightarrow \pi^{-1}U \cong Y/\mathbb{C}^\star$.

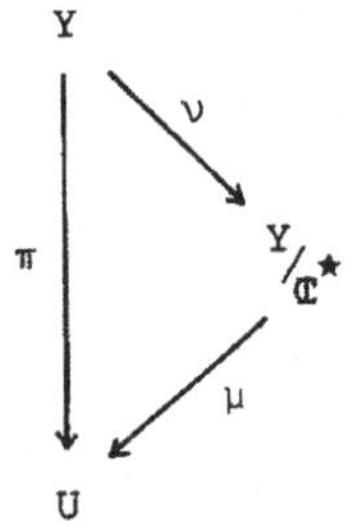

The Riemanian connection on P induces a connection on the locally isomorphic bundle $\tilde{P}$, and so ones on the vector bundles $\underline{V}_+, \underline{V}_-$. Covariant differentiation on $\underline{V}_+, \underline{V}_-$ and the tangent bundle $\underline{T} \cong \underline{V}_+ \otimes \underline{V}_-$ are related by the derivation law

$$\nabla_X(u \otimes v) = \nabla_X u \otimes v + u \otimes \nabla_X v, \tag{8.1}$$

where $u \in \Gamma(\underline{V}_+)$, $v \in \Gamma(\underline{V}_-)$.

Let $\{u^0, u^1\}$ be a basis of $\underline{V}_+$ over U, with dual basis $\{\lambda_0, \lambda_1\}$ regarded as functions on Y, and suppose that $\nabla u^i = u^j \otimes \omega_j^i$. By (4.5) the subbundle F of $T^\star Y$ generated by the real and imaginary components of $\omega_i = d\lambda_i + \lambda_j \mu^\star \omega_i^j$, $i = 0,1$, depends only on ∇ and not on the basis, so there is a natural decomposition

(8.2) $$T^{\star}Y = F \oplus \mu^{\star}T^{\star}M.$$

For any $u \in Y$, the complex line $\mathbb{C}u \in \mathbb{P}(\underline{V}_+)$ determines an almost complex structure I_u on T_xM by corollary 7.4. Indeed using (7.3) we can take I_u to be the almost complex structure whose space of (1,0)-forms is

$$\mathbb{C}u \otimes \underline{V}_- \subset \underline{V}_+ \otimes \underline{V}_- \cong \underline{T}^{\star}.$$

Consequently (8.2) gives rise to an almost complex structure on Y whose distribution of (1,0) forms is given by

(8.3) $$\Lambda_u^{1,0}Y = \{\omega_1,\omega_2\} \oplus \{\mu^{\star}(u \otimes v) : v \in \underline{V}_-\}.$$

By the Newlander-Nirenberg theorem, Y is a complex manifold if $d\alpha$ has no (0,2)-component for any $\alpha \in \Gamma(\Lambda^{1,0}Y)$. First consider the (1,0)-form $\mu^{\star}(u \otimes v) = \lambda_k \mu^{\star}(u^k \otimes v)$ on Y (here we have used $\mu^{\star}$ in a pointwise sense, but in future we omit it). Using (8.1) and the fact that the Riemannian connection has torsion $\tau = d + a\nabla$ zero,

$$\begin{aligned} d(u \otimes v) &= d\lambda_k \wedge (u^k \otimes v) - \lambda_k a\nabla(u^k \otimes v) \\ &= -a[(u^k \otimes v) \otimes d\lambda_k + u \otimes \nabla v + \lambda_k(u^j \otimes v) \otimes \omega_j^k] \\ &= -a[u \otimes \nabla v + (u^j \otimes v) \otimes \omega_j] \end{aligned}$$

which has no (0,2)-component by (8.3). Second, (4.6) gives

$$d\omega_i = \omega_j \wedge \sigma_i^j + \lambda_j \Omega_i^j$$

where Ω_i^j are the components of the curvature $\Omega \in \operatorname{End} \underline{V}_+ \otimes \Lambda^2\underline{T}^{\star}$ of $\underline{V}_+$. Since $\Omega(u) = \Omega(\lambda_j u^j) = u^i \otimes \lambda_j \Omega_i^j$, we need

<u>Lemma 8.2</u> For any $u \in Y$, $\Omega(u)$ has no (0,2)-component relative to the almost complex structure I_u.

<u>Proof</u>. It follows from (8.1) that the curvature of the tangent bundle $\underline{T}$ is essentially the sum of the curvatures of $\underline{V}_+$ and $\underline{V}_-$. In other words using

$$(8.4) \qquad \Lambda^2_+ \cong S^2V_+ \hookrightarrow V_+ \otimes V_+ \underset{\varepsilon}{\cong} V_+^\star \otimes V_+ \cong \text{End}\, V_+,$$

Ω is that half of the Riemannian curvature lying in $\underline{\Lambda}^2_+ \otimes \Lambda^2\underline{T}^\star$. Fix $u \in Y$, $\pi(u) = x$, and choose an oriented orthonormal basis $\{e^1, e^2, e^3, e^4\}$ of $T_x^\star M$ so that $\gamma = (e^1 + ie^2) \wedge (e^3 + ie^4)$ is a (2,0)-form relative to I_u, as in the proof of proposition 7.1. Then in the notation of theorem 5.1,

$$\Omega = a_{ij}\phi^i \otimes \phi^j + b_{ij}\phi^i \otimes \psi^j.$$

But $W_+ \equiv 0$ means $a_{ij} = t\delta_{ij}$, giving

$$\Omega = \tfrac{1}{2}t(2\phi^1 \otimes \phi^1 + \gamma \otimes \bar{\gamma} + \bar{\gamma} \otimes \gamma) + b_{ij}\phi^i \otimes \psi^j.$$

It suffices to show that $\frac{1}{2}t\gamma(u) \otimes \bar{\gamma} = 0$. Now $\gamma \in \Lambda^2(\mathbb{C}u \otimes \underline{V}_-)$, so in terms of (8.4), γ corresponds to $cu \otimes u \in S^2\underline{V}_+$ for some $c \in \mathbb{C}$, and as an endomorphism of $\underline{V}_+$,

$$\gamma(u') = c\,\varepsilon(u',u)u,$$

where ε is a skew form. Thus $\gamma(u) = 0$ as required. ∎

We have proved that Y is a complex manifold. Observe finally that the almost complex structure on Y is invariant by the group $\mathbb{C}^\star$ of non-zero complex scalars, which thus acts holomorphically. Therefore the quotient $Y/\mathbb{C}^\star$ admits a complex structure independent of the choice of lifting $\tilde{P}$, and Z is a complex manifold. ∎

Two observations concerning the complex structure on Z defined by (8.3) need to be made immediately. First note that even if M happens to be a complex manifold, the projection $\pi : Z \longrightarrow M$ can never be holomorphic. Second, the inclusion $i : \pi^{-1}(x) \hookrightarrow Z$ embeds each fibre $\mathbb{C}P^1$ holomorphically in Z. These two points illustrate a characteristic property of the "twistor space" Z : although it is the total space of a non-holomorphic fibration, the fibres are nevertheless complex submanifolds.

Examples. 1. A manifold certainly satisfying the hypothesis of theorem 8.1 is Euclidean space $\mathbb{R}^4$ with its standard flat Riemannian metric. The twistor space Z is the product $\mathbb{R}^4 \times \mathbb{C}P^1$, but not complex-analytically. However the projection $Z \longrightarrow \mathbb{C}P^1$ is holomorphic, and identifies Z with the total space of the holomorphic normal bundle of any fibre $\pi^{-1}(x) \cong \mathbb{C}P^1$. In the next section we shall see that this is $\zeta^{-1} \oplus \zeta^{-1}$, where ζ is the tautologous holomorphic line bundle over $\mathbb{C}P^1$. Now the torus $\mathbb{R}^4/_\Gamma \approx S^1 \times S^1 \times S^1 \times S^1$ inherits a flat metric from $\mathbb{R}^4$. This example is more interesting as the twistor space is a compact complex manifold that does not admit a Kähler metric, and was studied by Blanchard [B].

2. The twistor space Z will fibre holomorphically over $\mathbb{C}P^1$ whenever the vector bundle $\underline{\Lambda}^2_+$ with its induced connection is flat, which means that M is anti-self-dual and Ricci flat. A less trivial example of this phenomenon occurs when M is a Kähler K3 surface with the Calabi-Yau metric. Each point of $\mathbb{C}P^1$ corresponds to a covariant constant section of $\mathbf{S}(\underline{\Lambda}^2_+) \cong Z$, which is therefore the fundamental 2-form of a Kähler metric on M. The complex structures I,J,K associated to any three such orthogonal sections satisfy the quaternionic identity

$$IJ = -JI = K.$$

In fact the choice of I,J,K reduces the structure group $SO(4)$ of M to the subgroup $Sp(1) \subset \mathrm{Aut}(\underline{\Lambda}^2_-)$. Note that the canonical bundle $\kappa = \underline{\Lambda}^{2,0}$ determined by any one of the complex structures $aI + bJ + cK$, $a^2 + b^2 + c^2 = 1$, is necessarily trivial.

A simpler example of a complex manifold with trivial canonical bundle is the total space M of the holomorphic cotangent bundle $\pi : \Lambda^{1,0}\mathbb{C}P^1 \longrightarrow \mathbb{C}P^1$. For there is a short exact sequence

$$0 \longrightarrow \pi^\star\Lambda^{1,0}\mathbb{C}P^1 \longrightarrow \Lambda^{1,0}M \longrightarrow Q \longrightarrow 0$$

of vector bundles on M in which the quotient Q consists of $(1,0)$-forms on the fibres. Thus $Q \cong (\pi^\star\Lambda^{1,0}\mathbb{C}P^1)^\star$ and

$\kappa = \Lambda^{2,0}M \cong \pi^{\star}\Lambda^{1,0}\mathbb{C}P^1 \otimes Q$ is trivial. Calabi $[C_2]$ has shown that M also admits an anti-self-dual Ricci-flat metric; he calls such metrics "hyper-Kähler" because of the $\mathbb{C}P^1$-worth of Kähler structures. In this case I may be chosen to be the standard complex structure of M, whereas J and K interchange vertical and horizontal directions. Now $\Lambda^{1,0}\mathbb{C}P^1 \cong \zeta^2$ (see lemma 9.1), and the S^1-bundle inside ζ^2 is diffeomorphic to the quotient $S^3/\mathbb{Z}_2$, so that at infinity M looks like $\mathbb{R} \times S^3/\mathbb{Z}_2$. Hitchin $[H_2]$ has generalized Calabi's example by finding hyper-Kähler manifolds that behave like $\mathbb{R}^3 \times S^3/\mathbb{Z}_k$ at infinity, for all $k \geq 2$.

3. Any conformally flat manifold is anti-self-dual. Identify $\mathbb{R}^4$ with the quaternions $\mathbb{H}$ and let $\mathbb{Z}$ denote the cyclic group generated by right multiplication by a quaternion q with $|q| > 1$. Then $M = \mathbb{R}^4\backslash 0/\mathbb{Z} \approx S^1 \times S^3$ has a conformally flat metric. Although $\underline{\Lambda}^2_-$ is not flat with respect to the Riemannian connection, it does have a flat $GL(1,\mathbb{H})$-structure, and the twistor space $Z \approx S^1 \times S^2 \times S^3$ again fibres holomorphically over $\mathbb{C}P^1 \cong S^2$. For more details see [So]. Unlike the torus, M cannot admit a Kähler metric because $b_2 = 0$ or because b_1 is odd; each point of $\mathbb{C}P^1$ endows M with the complex structure of a Hopf manifold. It follows that Z is also non-Kähler; in fact as we shall see later the instances of Kähler twistor spaces are rare $[H_4]$.

Now consider $S^4 \cong \mathbb{H}P^1$ with its standard conformally flat metric. By proposition 3.1, $\underline{V}_+$ is the tautologous quaternionic line bundle over $\mathbb{H}P^1$ whose total space can be identified with $\mathbb{H}^2$ with its origin blown up (replaced by the zero section $\underline{0}$ of $\underline{V}_+$). Then $Y = \underline{V}_+\backslash\underline{0} \cong \mathbb{H}^2\backslash 0$, and

$$Z = \mathbb{P}_{\mathbb{C}}(\underline{V}_+) \cong \mathbb{P}_{\mathbb{C}}(\mathbb{H}^2) \cong \mathbb{C}P^3.$$

It is not hard to check that the complex structure of the twistor space Z coincides with the standard complex structure of $\mathbb{C}P^3$. Alternatively,

$$Z = P \times_{Sp(1)} \frac{Sp(1)}{U(2)} \cong \frac{Sp(2)}{Sp(1) \times U(1)} \cong \frac{SO(5)}{U(2)}.$$

In these homogeneous descriptions of $\mathbb{C}P^3$, the isotropy representation is reducible, corresponding to the vertical and horizontal directions in Z.

Identifying $\mathbb{R}^4$ with S^4 minus a point, the twistor space of $\mathbb{R}^4$ becomes $\mathbb{C}P^3$ minus a $\mathbb{C}P^1$. This is consistent because $\zeta^{-1} \oplus \zeta^{-1}$ is the holomorphic normal bundle of $\mathbb{C}P^1$ in $\mathbb{C}P^3$. However we have anticipated a later result here, namely the conformal invariance of the twistor space construction.

4. The complex projective plane with its usual orientation cannot admit an anti-self-dual metric for topological reasons, for its signature $\tau = 1$ can be expressed as an integral of the quantity $|W_+|^2 - |W_-|^2$ [H_1]. Indeed the standard Kähler metric on $\mathbb{C}P^2$ is self-dual (theorem 7.2). This presents no problem, because reversing the orientation of M in the statement of theorem 8.1 gives that the total space of $Z_- = \mathbb{P}_{\mathbb{C}}(\underline{V}_-)$ is a complex manifold. But using (7.3),

$$\mathbb{P}_{\mathbb{C}}(\underline{V}_-) = \mathbb{P}_{\mathbb{C}}(\underline{L} \otimes \underline{V}_-) \cong \mathbb{P}_{\mathbb{C}}(\underline{T}^{1,0})$$

is the projective holomorphic tangent bundle. A point of the latter determines a complex projective line in $\mathbb{C}P^2$, and Z_- is the flag manifold

$$\{(V_1, V_2) : 0 \in V_1 \subset V_2 \subset \mathbb{C}^3,\ \dim V_i = i\}.$$

In homogeneous terms, $Z_- \longrightarrow \mathbb{C}P^2$ is simply a fibration

$$\frac{U(3)}{U(1) \times U(1) \times U(1)} \longrightarrow \frac{U(3)}{U(2) \times U(1)},$$

though not the standard holomorphic one.

Theorem 8.1 has various analogues in higher dimensions. The most obvious concerns the total space of the bundle Z over an oriented Riemannian 2n-manifold (see the end of section 7); this has a natural almost complex structure for all n. As in the 4-dimensional case, the essential point is to use the Riemannian connection on M to split each tangent space $T_z Z$ into horizontal and vertical subspaces. The point z itself defines an almost complex structure on the horizontal subspace, whereas the vertical subspace, by definition tangent to the fibre, inherits a complex structure from that of the Hermitian symmetric space $SO(2n)/U(n)$. Integrability of the direct sum of these almost complex structures again involves the Weyl tensor W, but since the latter is irreducible for $n \geq 3$ one actually needs $W \equiv 0$, i.e. M conformally flat, to ensure that Z be a complex manifold.

In order to obtain a twistor space with less restrictive integrability conditions, it is necessary to consider an associated bundle with a "smaller" fibre. In this sense an appropriate generalization consists of taking M to be a quaternionic Kähler manifold. This is a 4n-dimensional Riemannian manifold whose holonomy lies in a certain subgroup $Sp(n)Sp(1)$ of $SO(4n)$, and admits locally a triple I,J,K of almost complex structures satisfying $IJ = -JI = K$, whose fundamental 2-forms generate a subbundle $B \subset \Lambda^2 T^* M$. In particular for $n = 1$ this is valid for an arbitrary oriented Riemannian 4-manifold and $B = \underline{\Lambda}^2_+$ (cf. example 2 above). In general each point of the 2-sphere bundle $\mathbf{S}(B)$ has the form $aI + bJ + cK$, $a^2 + b^2 + c^2 = 1$, and is an almost complex structure on the tangent space of M below. One can now proceed as before, adding almost complex structures defined on the vertical and horizontal subspaces. For $n \geq 2$, properties of the curvature arising from the holonomy condition guarantee that $\mathbf{S}(B)$ is always a complex manifold [S].

The above examples fit into a more general construction of twistor spaces that has been described by Bérard Bergery and Ochiai. Suppose that G is a closed subgroup of $GL(2n,\mathbb{R})$, and C a G-invariant complex submanifold of the complex homogeneous space $GL(2n,\mathbb{R})/GL(n,\mathbb{C})$ of all almost complex structures on $\mathbb{R}^{2n}$. Given a manifold M with a G-structure, i.e. a subbundle P of the principal bundle of frames, the associated bundle $P \times_G C$ parametrizes a collection of almost complex structures on M. A connection on P again gives rise to an almost complex structure on the total space, which will be a complex manifold provided the torsion and curvature satisfy suitable conditions [BO]. When G is a subgroup of $SO(2n)$, a candidate for P is the holonomy bundle of a Riemannian manifold M with holonomy group equal to G. In this case the Riemannian connection defined on P will automatically satisfy the torsion condition for the integrability of $P \times_G C$.

9. THE NORMAL BUNDLE

Let M be an oriented Riemannian 4-manifold with $W_+ \equiv 0$, so that the twistor space Z exists as a complex manifold. We have already remarked that each fibre of Z is embedded as a complex projective line, so at this point we recall some elementary facts relating to the holomorphic structure of $\mathbb{C}P^1$ in order to estabish subsequent notation. Let $\nu : \mathbb{C}^2\setminus 0 \longrightarrow \mathbb{C}^2\setminus 0 / \mathbb{C}^\star = \mathbb{C}P^1$ be the projection. Then the fibre of the tautologous holomorphic line bundle ζ at $z = \nu(\lambda_0,\lambda_1) \in \mathbb{C}P^1$ is

$$\zeta_z = \{(\lambda_0 a, \lambda_1 a) : a \in \mathbb{C}\} \subset \mathbb{C}^2.$$

<u>Lemma 9.1</u> The holomorphic tangent bundle (i.e. the bundle of complex (1,0)-vectors) of $\mathbb{C}P^1$ is given by

$$T^{1,0}\mathbb{C}P^1 \cong \zeta^{-2} = \zeta^\star \otimes \zeta^\star.$$

<u>Proof</u>. First one defines a holomorphic short exact sequence

$$(9.1) \qquad 0 \longrightarrow \zeta \longrightarrow \mathbb{C}P^1 \times \mathbb{C}^2 \xrightarrow{p} \zeta \otimes T^{1,0}\mathbb{C}P^1 \longrightarrow 0.$$

The inclusion of ζ in the trivial bundle arises from the definition of ζ. Fix $z \in \mathbb{C}P^1$ together with a representative $u \in \mathbb{C}^2\setminus 0$ with $z = \nu(u)$. Then

$$\mathbb{C}^2 \cong T_u^{1,0}(\mathbb{C}^2\setminus 0) \xrightarrow{\nu_\star} T_z^{1,0}\mathbb{C}P^1,$$

and given $v \in \mathbb{C}^2$, set $p(z,v) = u \otimes \nu_\star v$. It is easy to see that this is independent of the choice of u, and to check exactness. The lemma now follows from the isomorphism

$$\mathbb{C}P^1 \times \Lambda^2\mathbb{C}^2 \cong \zeta^2 \otimes T^{1,0}\mathbb{C}P^1$$

induced from (9.1).

Incidentally there is now a holomorphic exact sequence

$$0 \longrightarrow \zeta \longrightarrow \mathbb{C}P^1 \times \mathbb{C}^2 \longrightarrow \zeta^{-1} \longrightarrow 0,$$

and topologically $\zeta \oplus \zeta^{-1}$ is a trivial bundle. This can be expressed formally by the equation $(\zeta - 1)^2 = 0$. ∎

The tautologous bundle ζ has obvious local holomorphic sections, namely

$$\text{(9.2)} \quad \begin{aligned} s_0 &= (1, \lambda_1\lambda_0^{-1}) \quad \text{on} \quad U_0 = \mathbb{C}P^1\backslash[0,1] \\ s_1 &= (\lambda_0\lambda_1^{-1}, 1) \quad \text{on} \quad U_1 = \mathbb{C}P^1\backslash[1,0]. \end{aligned}$$

Define dual sections $t_i \in \Gamma(U_i, \zeta^{-1})$ by $t_i(s_i) = 1$. On $U_0 \cap U_1$, whereas $\lambda_0 s_0 = \lambda_1 s_1$, we have $\lambda_0^{-1} t_0 = \lambda_1^{-1} t_1$. Any linear form $f = a_0\lambda_0 + a_1\lambda_1$, $a_i \in \mathbb{C}$, then determines a global holomorphic section t_f of ζ^{-1} by setting

$$t_f = f\lambda_i^{-1} t_i \quad \text{on} \quad U_i.$$

Observe that the pullback $\nu^*\zeta$ has a canonical section ℓ, with respect to which $\nu^* t_f = f\ell^{-1}$. On the other hand, given a holomorphic section t of ζ^{-k}, $\nu^* t = f\ell^{-k}$ where f is a holomorphic function on $\mathbb{C}^2\backslash 0$, homogeneous of degree k. By Hartogs's theorem, f extends holomorphically to $\mathbb{C}^2$; when $k = 1$ this forces f to be linear and $t = t_f$, whereas in general

<u>Lemma 9.2</u> The space of holomorphic sections of ζ^{-k} is given by

$$H^0(\mathbb{C}P^1, \zeta^{-k}) \cong \begin{cases} 0, & k < 0 \\ S^k(\mathbb{C}^2)^* \cong \text{homogeneous polynomials of degree } k \text{ in } \lambda_0, \lambda_1, & k \geq 0. \end{cases}$$

Returning to the fibration $\pi : Z \longrightarrow M$, let F denote a fixed fibre $\pi^{-1}(x) \cong \mathbb{C}P^1$. As in the proof of theorem 8.1, define the vector bundles $\underline{V}_{\pm}$ over a neighbourhood U of x, and put $Y = \underline{V}_{+}\backslash \underline{0}$. Then $\pi^{-1}U \cong Y/\mathbb{C}^{\star}$ has a tautologous holomorphic line bundle $\zeta \subset \pi^{\star}\underline{V}_{+}$ whose fibre at $z = \nu(u)$, $u \in Y$, is simply $\zeta_z = \mathbb{C}u$. As the notation suggests, the restriction $\zeta|_F$ of ζ to the fibre is none other than the tautologous bundle over $\mathbb{CP}^1$ considered above. Although ζ is only defined on $\pi^{-1}U$, by lemma 9.1 ζ^{-2} is isomorphic to the bundle of complex (1,0)-vectors tangent to the fibres, and is globally defined on Z.

The <u>holomorphic normal bundle</u> N of F in Z is by definition the quotient in the short exact sequence

$$0 \longrightarrow T^{1,0}F \xrightarrow{i_\star} T^{1,0}Z|_F \longrightarrow N \longrightarrow 0$$

<u>Proposition 9.3</u> $N \cong \zeta^{-1} \otimes (\underline{V}_-)_x \cong \zeta^{-1} \oplus \zeta^{-1}$.

<u>Proof</u>. The dual $N^\star$ of N, or conormal bundle, is the kernel in the exact sequence

$$0 \longrightarrow N^\star \xrightarrow{q} \Lambda^{1,0}Z|_F \xrightarrow{i^\star} \Lambda^{1,0}F \longrightarrow 0. \tag{9.3}$$

The underlying real conormal bundle is none other than the trivial bundle $\pi^\star T_xM$ over F, and from (8.3),

$$N^\star_z = \pi^\star(\mathbb{C}u \otimes \underline{V}_-)_x = \zeta_z \otimes (\underline{V}_-)_x,$$

where $z = \nu(u)$, $u \in (\underline{V}_+)_x\backslash 0$. Thus over F we have an isomorphism

$$N^\star \cong \zeta \otimes (\underline{V}_-)_x$$

of complex vector bundles, which we shall show shortly preserves the respective holomorphic structures. Taking duals and recalling that $\underline{V}_\pm \cong \underline{V}^\star_\pm$ then gives the proposition.

Take a basis $\{u^0, u^1\}$ of $\underline{V}_+$ over U such that $\nabla u^i|_x = 0$, and suppose that λ_0, λ_1 are the corresponding coordinates on the total space of $\underline{V}_+$. Then $\lambda = \lambda_0\lambda_1^{-1}$ is a local coordinate on Z whose restriction to F is holomorphic. Let s_1 denote the local holomorphic section of $\zeta|_F$ defined as in (9.2) using the coordinates λ_0, λ_1. Replacing $N^\star$ in (9.3) by $\zeta \otimes (\underline{V}_-)_x$, it is enough to show that

$$s = q(s_1 \otimes v) = \lambda\pi^{\star}(u^0 \otimes v) + \pi^{\star}(u^1 \otimes v)$$

is a holomorphic section of $\Lambda^{1,0}Z|_F$ for any $v \in (\underline{V}_-)_x$. This is true iff $i^{\star}(\bar{\partial}\tilde{s}|_F) = 0$, where $\tilde{s} \in \Gamma(\pi^{-1}U, \Lambda^{1,0}Z)$ is any smooth (1,0)-form extending s, and here $i^{\star}$ is the natural projection $\Lambda^{1,1}Z|_F \longrightarrow \Lambda^{0,1}F \otimes (\Lambda^{1,0}Z|_F)$. Define $\tilde{s}$ by extending v to a section of $\underline{V}_-$ over U with $\nabla v|_x = 0$. Then

$$\bar{\partial}s|_F = \bar{\partial}\lambda \wedge \pi^{\star}(u^0 \otimes v),$$

and $i^{\star}(\bar{\partial}\tilde{s}|_F) = 0$ because $\bar{\partial}\lambda|_F = \bar{\partial}(\lambda|_F) = 0$. ∎

We can now rewrite (9.3) as

$$0 \longrightarrow \zeta \oplus \zeta \longrightarrow \Lambda^{1,0}Z|_F \longrightarrow \zeta^2 \longrightarrow 0.$$

Suppose that f is a holomorphic function defined globally on Z. Then the restriction α of df to any fibre F is a holomorphic section of $\Lambda^{1,0}Z|_F$. Since neither ζ^2 nor $\zeta \oplus \zeta$ admit holomorphic sections, $\alpha = 0$. Thus $df = 0$ on Z, and f is a constant. Using in addition the 2nd and 3rd exterior powers

$$0 \longrightarrow \Lambda^2 N^{\star} \longrightarrow \Lambda^{2,0}Z|_F \longrightarrow N^{\star} \otimes \Lambda^{1,0}F \longrightarrow 0$$

(9.4) $$0 \longrightarrow \Lambda^{3,0}Z|_F \longrightarrow \Lambda^2 N^{\star} \otimes \Lambda^{1,0}F \longrightarrow 0$$

of (9.3) as in $[H_2]$ yields

<u>Theorem 9.4</u> The twistor space Z has no non-constant holomorphic functions, no holomorphic r-forms, $r \geq 1$, and its canonical bundle is given by $\kappa \cong \zeta^4$.

<u>Proof</u>. It remains to establish the last statement, but from (9.4) there is certainly an isomorphism

$$\kappa = \Lambda^{3,0}Z \cong \zeta^4$$

of smooth vector bundles. The nowhere-vanishing section σ of

$\kappa \otimes \zeta^{-4}$ defining this isomorphism can be expressed locally as follows. Take coordinates λ_0, λ_1, $\lambda = \lambda_0\lambda_1^{-1}$ corresponding to some local basis $\{u^0, u^1\}$ of $\underline{V}_+$ as usual, and let ω_0, ω_1 be the associated forms on the total space of $\underline{V}_+$ as in the proof of theorem 8.1. The (1,0)-form

$$\omega = \lambda_1^{-2}(\lambda_1\omega_0 - \lambda_0\omega_1)$$

is well-defined on an open set of Z, whereas

$$\pi^\star(\lambda u_0 + u_1)^2 \in \pi^\star S^2\underline{V}_+ \cong \pi^\star \underline{\Lambda}_+^2$$

is a (2,0)-form complementary to ω. Then

$$\sigma = [\omega \wedge \pi^\star(\lambda u_0 + u_1)^2] \otimes t_1^{\,4},$$

and its invariant nature can be seen by pulling back to $\underline{V}_+ \backslash \underline{0}$, where at a point u we have

$$\nu^\star\sigma|_u = (\lambda_1\omega_0 - \lambda_0\omega_1) \wedge \pi^\star(u^2).$$

The fact that the independently defined holomorphic structures of κ and ζ^4 coincide now follows by showing that $\bar\partial(\nu^\star\sigma) = 0$. We omit this computation since similar ones will be carried out in the next section when the exact sequence (9.3) is examined globally on Z. ∎

Fix a fibre $F = \pi^{-1}(x)$ of Z, and let $H^i(F,N) = H^i(F,\mathcal{O}(N))$ denote cohomology with coefficients in the sheaf of germs of local holomorphic sections of N. By the deformation theory of Kodaira [Ko], the vanishing of

$$H^1(F,N) \cong H^0(F,N^\star \otimes \zeta^2)^\star \cong H^0(\mathbb{CP}^1, \zeta^3 \oplus \zeta^3)^\star$$

implies that F belongs to a complex family of projective lines in Z whose tangent space at F is isomorphic to $H^0(F,N) \cong \mathbb{C}^4$. This family is a complexification of (an open set of) M because the fibres of Z constitute a real subfamily. More precisely, let ι denote the involution of $Z \cong \mathbb{S}(\underline{\Lambda}_+^2)$ induced by multiplication by -1 in $\underline{\Lambda}_+^2$ that sends a point on the sphere to its antipode. In terms of the description $Z = \mathbb{P}(\underline{V}_+)$, ι is determined by the quaternionic structure map of

$\underline{V}_+$:

$$\iota(\nu(u)) = \nu(ju), \quad u \in \underline{V}_+\backslash\underline{0}.$$

The differential $d\iota$ acts antilinearly on the tangent bundle of Z, and ι defines a real structure on Z with no fixed points, but leaving invariant the fibres.

The following result illustrates the above remarks infinitesimally, and shows that the conformal structure of M can be extracted from the holomorphic structure of Z.

Proposition 9.5 The space $H^0(F,N)$ of holomorphic sections of N over $F = \pi^{-1}(x)$ is naturally isomorphic to T_xM and the kernel of the obvious mapping

$$e : S^2(H^0(F,N)) \longrightarrow H^0(F,S^2N)$$

is spanned by the Riemannian metric.

Proof. By proposition 9.3 and lemma 9.2,

$$\begin{aligned} H^0(F,N) &\cong H^0(F,\zeta^{-1}) \otimes (\underline{V}_-)_x \\ &\cong (\underline{V}_+ \otimes \underline{V}_-)_x \\ &\cong (T_xM)_c. \end{aligned}$$

We must define a real structure map (see section 1) on $H^0(F,N)$ so that the above becomes an isomorphism of real vector spaces. For this purpose note that the differential $d\iota$ induces a vector space isomorphism

$$d\iota : N_z \xrightarrow{\cong} \overline{N_{\iota z}}.$$

Complex conjugation in $(T_xM)_c$ then corresponds to the involution $t \longmapsto \bar{t}$ of $H^0(F,N)$ given by

$$\bar{t}(\iota z) = \overline{d\iota(t(z))}.$$

Finally,

$$S^2N \cong S^2(\zeta^{-1} \otimes (\underline{V}_-)_x) \cong \zeta^{-2} \otimes (S^2\underline{V}_-)_x$$

and

$$\begin{aligned} H^0(F,S^2N) &\cong H^0(F,\zeta^{-2}) \otimes (S^2\underline{V}_-)_x \\ &\cong (S^2\underline{V}_+ \otimes S^2\underline{V}_-)_x \\ &\cong S^2_0T_xM \end{aligned}$$

by proposition 2.2. It can now be checked that in these terms, e is the projection

$$S^2T_xM \longrightarrow S^2_0T_xM$$

determined by the conformal class of the metric. ∎

Instead of taking symmetric powers of the holomorphic normal bundle N over $F = \pi^{-1}(x)$, consider the exterior power

$$\Lambda^2N \cong \Lambda^2(\zeta \otimes (\underline{V}_-)_x) \cong \zeta^2 \otimes (\Lambda^2\underline{V}_-)_x \cong \zeta^2.$$

Then

$$\begin{aligned} H^0(F,\Lambda^2N) &\cong H^0(F,\zeta^2) \\ &\cong (S^2\underline{V}_+)_x \\ &\cong (\underline{\Lambda}^2_+)_x, \end{aligned}$$

and so the holomorphic structure determines the +1-eigenspace of the $\star$-operator on $\Lambda^2T^\star_xM$. Of course the latter is conformally invariant; indeed the decomposition

$$\Lambda^2T^\star M = \underline{\Lambda}^2_+ \oplus \underline{\Lambda}^2_-$$

actually determines the conformal structure of M.

<u>Example</u>. The complexification M^c of $M = S^4$ in the sense discussed above is the Grassmannian of complex 2-dimensional subspaces in $\mathbb{C}^4$.

An $Sp(2,\mathbb{C})$ structure on $\mathbb{C}^4$ determines a $Spin(4,\mathbb{C})$, i.e. an $SL(2,\mathbb{C}) \times SL(2,\mathbb{C})$ structure on the tangent bundle of M^c. Complex analytic spinor bundles $\underline{V}^c_{\pm}$ are then defined over M^c, and $\underline{V}_+$ can be identified with the tautologous bundle whose fibre over $x \in M^c$ consists of the corresponding 2-plane in $\mathbb{C}^4$. It follows that the complex projective line in $Z = \mathbb{C}P^3$ associated to x can be identified with the fibre of the projective bundle $Y = \mathbb{P}(\underline{V}^c_+)$. This gives rise to a diagram of fibrations

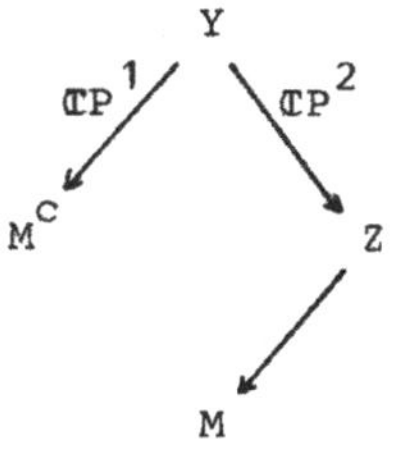

which can be used to study the twistor space Z [EPW]. The relevance of the conformal structure of M in this example will be discussed in section 12.

10. KÄHLER TWISTOR SPACES

The properties of the twistor space Z of an anti-self-dual Riemannian 4-manifold examined so far relate to the conformal structure of M. This section is devoted to a study of certain additional structures on Z associated specifically to the Riemannian metric of M. For instance the Riemannian connection determines a splitting of each tangent space of Z into a vertical and horizontal part; we shall see next that this is particularly significant when M is Einstein.

From lemma 9.1 and proposition 9.3, there exists a short exact sequence

$$(10.1) \qquad 0 \longrightarrow \zeta \otimes \pi^{\star}\underline{V}_{-} \xrightarrow{q} \Lambda^{1,0}Z \xrightarrow{r} \zeta^{2} \longrightarrow 0$$

of complex vector bundles defined globally on Z. Although its restriction (9.3) to each fibre of Z is holomorphic, (10.1) itself is in fact never a holomorphic sequence. However we do have the "reverse":

<u>Theorem 10.1</u> Let M be an anti-self-dual <u>Einstein</u> 4-manifold. Then on Z there exists a holomorphic short exact sequence

$$(10.2) \qquad 0 \longrightarrow \zeta^{2} \xrightarrow{s} \Lambda^{1,0}Z \longrightarrow \zeta \otimes \pi^{\star}\underline{V}_{-} \longrightarrow 0 .$$

<u>Proof</u>. The decompositions (8.2), (8.3) arising from the Riemannian connection of M give rise to a splitting of the sequence (10.1). For $\Lambda^{1,0}Z$ contains a distinguished 1-dimensional subbundle $\tilde{F}^{1,0}$ whose pullback $\nu^{\star}\tilde{F}^{1,0}$ lies in the bundle generated by ω_0, ω_1, using the notation of section 8. Locally $\tilde{F}^{1,0}$ is spanned by the form

$$\omega = \lambda_1^{-2}(\lambda_1\omega_0 - \lambda_0\omega_1) = d\lambda - \lambda^2\omega_1^0 + \omega_0^1 + \lambda(\omega_0^0 - \omega_1^1),$$

introduced in the proof of theorem 9.4 (on the extreme right $\pi^{\star}$ has been omitted). Then r induces an isomorphism $\tilde{F}^{1,0} \cong \zeta^2$ which maps the form ω to the section $s_1^{\,2}$ of ζ^2. The associated splitting s satisfying $rs = 1$ is represented by the ζ^{-2}-valued 1-form $\omega \otimes t_1^{\,2}$.

To prove that s is holomorphic it suffices to show that the exterior derivative of the invariant form

$$\tau = \nu^{\star}(\omega \otimes t_1^{\,2}) = \lambda_1\omega_0 - \lambda_0\omega_1$$

has type (2,0) on Y. Fix $u \in Y$ and suppose that the basis $\{u^0, u^1\}$ of $\underline{V}_+$ has been chosen so that

(10.3)

(i) it is compatible with the SU(2)-structure, i.e. $u^1 = ju^0$ and $\varepsilon(u^0,u^1) = 1$,

(ii) $u = u^1(x)$, i.e. u has coordinates $\lambda_0 = 0$, $\lambda_1 = 1$,

(iii) it is covariant constant at x, i.e. $\nabla u^i|_x = 0$.

Then

$$\begin{aligned} d\tau|_u &= d\omega_0 - 2d\lambda_0 \wedge d\lambda_1 \\ &= d(d\lambda_0 + \lambda_0\omega_0^0 + \lambda_1\omega_0^1) - 2d\lambda_0 \wedge d\lambda_1 \\ &= d\omega_0^1|_x - 2d\lambda_0 \wedge d\lambda_1 \\ &= \Omega_0^1 - 2d\lambda_0 \wedge d\lambda_1 \end{aligned}$$

where $\Omega(u^i) = u^j \otimes \Omega_j^i$ is the curvature of $\underline{V}_+$. From the proof of lemma 8.2,

$$\Omega = \tfrac{1}{2}t(2\phi^1 \otimes \phi^1 + \gamma \otimes \bar{\gamma} + \bar{\gamma} \otimes \gamma),$$

and in view of (i) we can identify up to constants

$$\phi^1 = u^0 \vee u^1, \qquad \bar{\gamma} = u^0 \otimes u^0, \qquad \gamma = u^1 \otimes u^1$$

It follows that $\Omega_0^1|_x = \frac{1}{2}t\gamma$, and $d\tau$ is a (2,0)-form as required. Having established that s is holomorphic, we can conclude that its quotient $\zeta \otimes \pi^{\star}\underline{V}_-$ inherits a holomorphic structure so as to render the corresponding short exact sequence (10.2) holomorphic. ■

Corollary 10.2 Under the hypotheses of theorem 10.1, the vector bundle $\pi^* \underline{V}_-$ defined over some open set $\pi^{-1}U \subset Z$ is holomorphic.

This result is connected to the fact that the vector bundle $\underline{V}_-$ is anti-self-dual, that is to say its curvature 2-forms belong to $\underline{\Lambda}^2_-$ (see section 5). Actually if E is any complex vector bundle over an anti-self-dual 4-manifold, with a connection whose curvature is a section of $\operatorname{End} E \otimes \underline{\Lambda}^2_-$, then the pullback $\pi^* E$ over the twistor space has a holomorphic structure [AHS]. This fact led to a classification of all such vector bundles over S^4 [ADHM]. It illustrates an important use of the twistor space, that of encoding certain real differential-geometric objects into holomorphic data.

Taking the 3rd exterior power of (10.2) gives immediately the isomorphism

$$\kappa = \Lambda^{3,0}Z \cong \zeta^4 \tag{10.4}$$

of holomorphic vector bundles that holds more generally (theorem 9.4). On the other hand, regarding $s \in H^0(Z, \Lambda^{1,0}Z \otimes \zeta^{-2})$ as a ζ^{-2}-valued 1-form, the quantity

$$s \wedge ds \in H^0(Z, \Lambda^{3,0}Z \otimes \zeta^{-4}) \cong \mathbb{C}$$

(theorem 9.4, Z connected) makes sense because it is independent of the local trivialization of ζ^{-2} chosen to compute ds. Since M is Einstein, its scalar curvature t is constant, and a computation shows that $s \wedge ds$ is nowhere zero provided $t \neq 0$. In this case s gives rise to a complex contact structure on Z, which on a complex $(2n+1)$-dimensional manifold is normally defined by an open covering $\{U_i\}$ and a collection $\{\alpha_i\}$ of holomorphic 1-forms such that $\alpha_i \wedge (d\alpha_i)^n$ is nowhere-zero on U_i, and $\alpha_i = f_{ij}\alpha_j$ on $U_i \cap U_j$. The nowhere-zero holomorphic functions f_{ij} are simply the transition functions of the line bundle ζ^2 [K]. The Einstein metric on M can also be extracted from the contact structure of Z [LB].

The line bundle ζ, defined on some $\pi^{-1}U = \nu(Y) \subset Z$, admits a natural unitary structure which we describe next. Any point $u \in Y$ is also a point of the fibre $(\nu^{*}\zeta)_u$, defining the canonical section of $\nu^{*}\zeta$. Now take a basis $\{u^0, u^1\}$ of $\underline{V}_+$ over U satisfying (10.3) (i), with corresponding coordinates λ_0, λ_1 on Y. If s is a local holomorphic section of ζ, $\nu^{*}s = f\ell$ for some local holomorphic function f on Y, homogeneous of degree -1. Then the norm of s at $z \in \pi^{-1}U$ is defined by

$$\|s\|_z = (|\lambda_0|^2 + |\lambda_1|^2)^{\frac{1}{2}}|f|, \tag{10.5}$$

where the right hand side is evaluated at any point of $\nu^{-1}(z)$.

<u>Theorem 10.3</u> Let M be an anti-self-dual Einstein 4-manifold with $t > 0$. Then its twistor space Z is a Kähler-Einstein manifold.

<u>Proof</u>. Our candidate for the Kähler metric is the closed (1,1)-form Θ defined on Z by setting

$$\Theta = -\bar{\partial}\partial \log \|s\|^2,$$

where s is any local non-zero holomorphic section of ζ. This is independent of the choice of s, because if $s' = fs$ is another holomorphic section,

$$\bar{\partial}\partial \log \|s'\|^2 - \bar{\partial}\partial \log \|s\|^2 = \bar{\partial}\partial(\log f + \log \bar{f}) = 0.$$

Moreover, because the line bundle $\zeta^{-2} \cong T^{1,0}(\text{fibres})$ is well-defined globally on Z, so is the form Θ. It suffices to show that the bilinear form $\Theta(IX,Y)$ is positive definite, where I is the almost complex structure of Z (see section 6).

Using (10.5),

$$\begin{aligned}\nu^{*}\Theta &= -\bar{\partial}\partial \log \nu^{*}\|s\|^2 \\ &= -d\partial \log (|\lambda_0|^2 + |\lambda_1|^2) \\ &= -d\left[\frac{\lambda_0\partial\bar{\lambda}_0 + \bar{\lambda}_0\partial\lambda_0 + \lambda_1\partial\bar{\lambda}_1 + \bar{\lambda}_1\partial\lambda_1}{|\lambda_0|^2 + |\lambda_1|^2}\right].\end{aligned}$$

Fix $u \in Y$ and suppose that $\{u^0, u^1\}$ is chosen to satisfy all the conditions of (10.3). Then at u, $\omega_i = d\lambda_i + \lambda_j \omega_i^j = d\lambda_i$ is a (1,0)-form, and

$$\nu^* \Theta|_u = d\lambda_0 \wedge d\bar{\lambda}_0 - d\partial\bar{\lambda}_1 - d\partial\lambda_1 .$$

Because the connection forms ω_i^j are skew-Hermitian,

$$\begin{aligned} d(\partial\lambda_1 + \partial\bar{\lambda}_1)|_u &= d(\omega_1 - (\omega_1^1 + \bar{\omega}_1^1)^{1,0}) \\ &= d\omega_1 \\ &= d\omega_1^1|_x \\ &= \Omega_1^1 . \end{aligned}$$

The function $\lambda = \lambda_0 \lambda_1^{-1}$ is well-defined near $z = \nu(u)$, and so

$$\Theta|_z = d\lambda \wedge d\bar{\lambda} - \Omega_1^1|_x .$$

From the proof of theorem 10.1, it follows that $\Omega_1^1|_x = \frac{1}{2} t\phi^1$, where ϕ^1 is the fundamental 2-form corresponding to the almost complex structure I_u on $T_x M$ defined by u. Consequently $\Theta(IX,Y)$ is a Riemannian metric, and Z is Kähler.

The holomorphic line bundle ζ has a unique connection compatible with both its holomorphic and unitary structures. This is defined by $\nabla s = s \otimes \partial \log \|s\|^2$, where s is any local non-zero holomorphic section, and its curvature is the 2-form $d\partial \log \|s\|^2 = -\Theta$. Because (10.4) identifies the holomorphic and unitary structures of ζ^4 with those of κ induced from the Kähler metric, the connection defined on ζ^4 by ∇ must coincide with the Kähler one on κ. The curvature of the latter is therefore -4Θ, but the associated bilinear form $-4\Theta(IX,Y)$ is known to equal the Ricci tensor (up to a constant) [C_1]. ■

In reality the hypotheses of theorem 10.3 are very restrictive, for its conclusion is the first step in the proof of the following deeper result.

Theorem 10.4 A complete, connected, anti-self-dual Einstein 4-manifold M with positive scalar curvature is isometric to S^4 or $\mathbb{C}P^2$ (the latter with its opposite orientation).

First Myers's theorem [KN; VIII, theorem 5.8] implies that M is necessarily compact, so its twistor space Z is a compact Kähler manifold. The theorem is then essentially a result of Hitchin [H_4] which asserts that under the last hypothesis, M must be conformally equivalent to S^4 or $\mathbb{C}P^2$. The equivalence at the Riemannian level can be deduced by exploiting the Einstein structures of Z and M [S]. See also Friedrich and Kurke [FK].

With the hypotheses of theorem 10.4 and the additional assumption that M is Spin, one can reach the conclusion that M is isometric to S^4 more readily. In this case the spinor bundles $\underline{V}_+$, $\underline{V}_-$ are defined over all of M, and the complex line bundle ζ over all of Z. The exact sequence (10.1) or (10.2) enables one to express the Chern classes of Z in terms of those of ζ and $\underline{V}_-$; in particular

$$c_1(Z) = -c_1(\kappa) = -c_1(\zeta^4) = -4c_1(\zeta).$$

Related results are valid on an arbitrary complex contact manifold [K]. Now $c_1(\zeta)$ is represented in deRham cohomology by $\frac{-i}{2\pi}\Theta$, where the curvature form $-\Theta$ of ζ is negative definite. Therefore the first Chern class of the compact Kähler manifold Z is 4 times a positive integral (1,1)-cohomology class.

A characterization of complex projective space by Kobayashi and Ochiai [KO] guarantees that Z is biholomorphically equivalent to $\mathbb{C}P^3$. The reason is that, using vanishing theorems, the cubic polynomial

$$P(k) = \chi(Z,\zeta^{-k}) = \sum_{i=1}^{3} (-1)^i \dim H^i(Z,\zeta^{-k})$$

satisfies $P(-1) = P(-2) = P(-3) = 0$, and $P(0) = 1$. The last equation gives the Todd genus of Z, and also follows from theorem 9.4. Then $P(k)$ coincides with the polynomial formed by taking Z to be $\mathbb{C}P^3$ and ζ the tautologous line bundle with first Chern class equal to -1. It is shown that $H^0(Z,\zeta^{-1}) \cong \mathbb{C}^4$ has no base points, and choosing a basis $\{s_i\}$ of sections of ζ^{-1},

$$x \longmapsto (s_1(x),\dots,s_4(x))$$

maps Z biholomorphically onto $\mathbb{C}P^3$. Since Z is known to be Kähler-Einstein, it must be isometric to $\mathbb{C}P^3$ [M], and a study of those isometries preserving the fibres gives $M \cong S^4$ as Riemannian manifolds.

A purely differential-geometric proof of theorem 10.4 is also available that with hindsight does not make explicit use of the complex manifold Z. The essential step uses the Atiyah-Singer index theorem [AS] to prove that the dimension of the isometry group of M satisfies

$$I = 10 - 2b_2^- \geqslant 4, \tag{10.6}$$

where $b_2^- = \dim\{\phi \in \Gamma(\underline{\Lambda}_-^2) : d\phi = 0\}$; see [S] for the methods involved. The cases $b_2^- = 3,2$ can be successively ruled out by using the fact that the isometry group leaves invariant the harmonic forms ϕ.

11. DIFFERENTIAL OPERATORS AND COHOMOLOGY

The results of section 9 can be extended to relate Dolbeault cohomology on the complex 3-dimensional twistor space to certain differential operators on the real 4-dimensional base manifold. Let $F = \pi^{-1}(x)$ be a fibre of the twistor space Z of an anti-self-dual Riemannian 4-manifold M. From section 9, the vector spaces $H^0(F,\Lambda^r N)$, $0 \leq r \leq 2$ have real structures, and are isomorphic to the fibres of the vector bundles $\underline{\Lambda}^0, \underline{\Lambda}^1, \underline{\Lambda}^2_+$ over M at x. These vector bundles form a subcomplex

$$0 \longrightarrow \underline{\Lambda}^0 \xrightarrow{D} \underline{\Lambda}^1 \xrightarrow{D} \underline{\Lambda}^2_+ \longrightarrow 0 \tag{11.1}$$

of the deRham complex of M in which the differential operator D is exterior differentiation, followed by a projection in the second case.

To relate this to Z we make use of the short exact sequence (10.1) of vector bundles over Z. The unitary structure of the line bundle ζ gives rise to an isomorphism $\bar{\zeta} \cong \zeta^{-1}$ of its conjugate with its dual, which in terms of (9.2) is given explicitly by

$$\begin{aligned} \bar{s}_0 &\longmapsto (1 + |\lambda|^{-2})t_0 \\ \bar{s}_1 &\longmapsto (1 + |\lambda|^2)t_1, \quad \lambda = \lambda_0\lambda_1^{-1}. \end{aligned} \tag{11.2}$$

On the other hand $\bar{\underline{V}}_- \cong \underline{V}_-^* \cong \underline{V}_-$, so the complex conjugate of (10.1) is

$$0 \longrightarrow N \longrightarrow \Lambda^{0,1}Z \longrightarrow \zeta^{-2} \longrightarrow 0$$

where $N = \zeta^{-1} \otimes \pi^*\underline{V}_-$ restricted to each fibre is the normal bundle. Taking second exterior powers gives another short exact sequence

$$0 \longrightarrow \Lambda^2 N \longrightarrow \Lambda^{0,2}Z \longrightarrow \zeta^{-2} \otimes N \longrightarrow 0.$$

Fix an open set $U \subset M$, and put

$$B^i = \Gamma(\pi^{-1}U, \Lambda^i N),$$

$$C^i = \Gamma(\pi^{-1}U, \Lambda^{0,i}Z),$$

$$D^i = \Gamma(\pi^{-1}U, \Lambda^{i-1}N \otimes \zeta^{-2}),$$

where as always Γ denotes smooth sections, and $\pi : Z \hookrightarrow M$ is the projection. Then there is a diagram

$$\text{(11.3)} \qquad \begin{array}{ccccccccc} & & & & D^1 & & D^2 & & D^3 & & \\ & & & & \uparrow\psi & & \uparrow\psi & & \uparrow\psi\ \wr\!\wr & & \\ 0 & \longrightarrow & C^0 & \xrightarrow{\bar\partial} & C^1 & \xrightarrow{\bar\partial} & C^2 & \xrightarrow{\bar\partial} & C^3 & \longrightarrow & 0 \\ & & \uparrow\phi\ \wr\!\wr & & \uparrow\phi & & \uparrow\phi & & & & \\ & & B^0 & & B^1 & & B^2 & & & & \end{array}$$

The vertical sequences are short exact, whereas the central horizontal row is the Dolbeault complex on $\pi^{-1}U$. Thus if $c \in C^i$, $\bar\partial c$ is simply the $(0,i+1)$-component of the exterior derivative dc.

Forgetting the interpretation of (11.3) for the moment, let us concentrate on the abstract homological algebra. Let A^i, E^{i+1} denote the kernel and cokernel respectively of the diagonal composition $\bar\partial' = \psi \circ \bar\partial \circ \phi$, so that there is an exact sequence

$$0 \longrightarrow A^i \longrightarrow B^i \xrightarrow{\bar\partial'} D^{i+1} \longrightarrow E^{i+1} \longrightarrow 0.$$

<u>Lemma 11.1</u> There are induced complexes

$$0 \longrightarrow A^0 \xrightarrow{D} A^1 \xrightarrow{D} A^2 \longrightarrow 0$$

$$0 \longrightarrow E^1 \xrightarrow{D} E^2 \xrightarrow{D} E^3 \longrightarrow 0$$

such that $A^\star \longrightarrow C^\star \longrightarrow E^\star$ are cochain mappings, and a long exact sequence of associated cohomology groups

$$\cdots \longrightarrow H^i(A) \longrightarrow H^i(C) \longrightarrow H^i(E) \xrightarrow{d} H^{i+1}(A) \longrightarrow \cdots$$

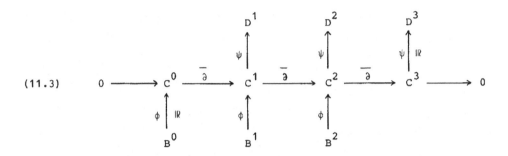

Proof. This consists of diagram chasing (alternatively the use of a spectral sequence). Suppose $a \in B^i$ satisfies $\bar\partial' a = 0$. Then $\psi\bar\partial\phi a = 0$, so $\bar\partial\phi a = \phi b$ for some $b \in B^{i+1}$. Moreover $\bar\partial' b = \psi\bar\partial\phi b = \psi\bar\partial\bar\partial\phi a = 0$, so $b \in A^{i+1}$. Put $D(a) = b$. The complex $(E^\star, D)$ is defined analogously. As for the long exact sequence we shall only bother to define the coboundary operator d. Suppose $e \in E^i$ satisfies $De = 0$. Choose $c \in C^i$ mapping to e; then $\psi\bar\partial c$ maps to $De = 0$ in E^{i+1}, and $\psi\bar\partial c = \bar\partial' b$ for some $b \in B^i$. Now $\psi\bar\partial(c - \phi b) = 0$, so $\bar\partial(c - \phi b) = a$ for some $a \in B^{i+1}$ satisfying $\bar\partial a = 0$, i.e. $a \in A^{i+1}$ and $Da = 0$. The assignment $e \longmapsto a$ induces the homomorphism $d : H^i(E) \longrightarrow H^{i+1}(A)$ which is defined independently of the choices involved. ∎

Lemma 11.2 The restriction of $\bar\partial' : B^i \longrightarrow D^{i+1}$ to a fibre $F = \pi^{-1}(x) \cong \mathbb{C}P^1$ is well-defined and coincides with the Dolbeault complex

$$0 \longrightarrow \Lambda^{0,0}F \otimes \Lambda^i N \xrightarrow{\bar\partial_F} \Lambda^{0,1}F \otimes \Lambda^i N \longrightarrow 0$$

on F with coefficents in the holomorphic vector bundle $\Lambda^i N$.

Proof. The value of $\bar\partial' b$, $b \in B^i$, at $z \in F$ depends only on the restriction of b to F. For if b vanishes on F, $b = rb_1$ for some function r on M with $r(x) = 0$ and $b_1 \in B^i$. Then $\bar\partial' b = \psi(\bar\partial r \wedge \phi b_1 + r\bar\partial\phi b_1) = 0$ on F.

Certainly $\bar\partial' = \psi\bar\partial = \bar\partial_F$ on F for $i = 0$; we tackle the case $i = 1$. Fix a point $z \in Z$. If $b \in B^1$ defines a holomorphic section of N over F, and f is a smooth function on Z, we must show that $\bar\partial'(fb) = \bar\partial_F f \otimes b$ at z. Having chosen a local basis $\{u^0, u^1\}$ of $\underline{V}_+$ as in (10.3) such that $z = \nu(u)$ has coordinate $\lambda = \lambda_0\lambda_1^{-1} = 0$, it suffices to take $b = t_1 \otimes \pi^\star v \in \zeta^{-1} \otimes \pi^\star \underline{V}_- \cong N$ with $\nabla v|_x = 0$. By (9.2), (11.2), omitting $\pi^\star$,

$$\phi(fb) = f(1 + |\lambda|^2)^{-1} j(\lambda u^0 + u^1) \otimes jv$$

$$= f(1 + |\lambda|^2)^{-1}(-u^0 + \bar\lambda u^1) \otimes jv$$

Thus

$$\bar{\partial}'(fb)|_z = \psi(d\bar{\lambda} \wedge (u^1 \otimes jv) + \bar{\partial}f \wedge \phi b) = \bar{\partial}_F f \otimes b,$$

since $u^1 \otimes jv$ is a (1,0)-form at z. ■

It now follows that A^i is the space of smooth sections of the vector bundle $x \longmapsto H^0(\pi^{-1}(x), \Lambda^i N)$, i.e.

$$A^0 = \Gamma(U, \underline{\Lambda}^0), \qquad A^1 = \Gamma(U, \underline{\Lambda}^1), \qquad A^2 = \Gamma(U, \underline{\Lambda}^2_+).$$

Furthermore it can be verified that the homomorphism $\phi : A^i \longrightarrow C^i$ is the obvious one which maps a form on U to the $(0,i)$-component of its pullback on $\pi^{-1}U$. This means that the operator $D : A^i \longrightarrow A^{i+1}$ of lemma 11.1 is induced from exterior differentiation on Z. Then using the fact that $\pi^{\star}\underline{\Lambda}^2_- \subset \Lambda^{1,1}Z$ (see proposition 7.1), the complex $(A^{\star}, D)$ is none other than (11.1).

At the sheaf level, the Dolbeault complex on a complex manifold is an acyclic resolution of the sheaf $\mathcal{O}$ of germs of holomorphic functions. The "deRham theorem" gives an isomorphism

$$H^i(C) = {}^{\ker \bar{\partial}}/_{\mathrm{im}\, \bar{\partial}} \cong H^i(\pi^{-1}U, \mathcal{O})$$

between the cohomology of $(C^{\star}, \bar{\partial})$ and the Cech cohomology groups of the sheaf $\mathcal{O}$. Moreover over a fibre $F \cong \mathbb{C}P^1$, $N \cong \zeta^{-1} \oplus \zeta^{-1}$ so

$$\operatorname{coker} \bar{\partial}_F \cong H^1(F, \Lambda^i N) \cong H^0(F, \Lambda^i N^{\star} \otimes \zeta^2)^{\star} = 0$$

by Serre duality. It follows that $E^i = 0$ for all i, and as a corollary of lemma 11.1 we have

<u>Theorem 11.3</u> For any open set $U \subset M$,

$$H^i(\pi^{-1}U, \mathcal{O}) \cong \begin{cases} H^i(A), & 0 \leq i \leq 2 \\ 0, & i = 3 \end{cases}.$$

We make two observations. First the above theorem is really a statement about <u>direct image sheaves</u>. The direct image $\pi^i_\star \mathcal{O}$ of $\mathcal{O}$ under π is the sheaf associated to the presheaf $U \longmapsto H^i(\pi^{-1}U,\mathcal{O})$ and is thus isomorphic to the cohomology <u>sheaf</u> of the complex of sheaves arising from (A^i,D). Second if $U = M$ is compact, the fact that $H^3(Z,\mathcal{O}) = 0$ also follows from Serre duality.

Clearly $H^0(A) \cong \mathbb{C}$ for any connected $U \subset M$; hence $H^0(\pi^{-1}U,\mathcal{O}) \cong \mathbb{C}$, a result also established in theorem 9.4. Unlike the deRham complex though, (11.1) does not possess the Poincaré lemma at the next step. For example on any neighbourhood U of flat space $\mathbb{R}^4$,

$$\alpha = x^1 dx^2 - x^3 dx^4$$

is a 1-form such that $0 \neq d\alpha \in \Gamma(U,\underline{\Lambda}^2_-)$ and defines a non-zero element of $H^1(A)$. More generally, if U admits a complex structure compatible with the metric and orientation, using proposition 7.1 (11.1) becomes

$$0 \longrightarrow \underline{\Lambda}^{0,0} \longrightarrow \underline{\Lambda}^{1,0} \oplus \underline{\Lambda}^{0,1} \longrightarrow \underline{\Lambda}^{2,0} \oplus \underline{\Lambda}^{0,2} \oplus \{F\} \longrightarrow 0.$$

Suppose given $\sigma \in \Gamma(U,\underline{\Lambda}^{1,0})$ with $D(\sigma + \bar{\sigma}) = 0$. Then $\bar{\partial}\bar{\sigma} = 0$, so by the Poincaré lemma for $\bar{\partial}$, $\bar{\sigma} = i\bar{\partial}f$ on some smaller $U' \subset U$. In addition $i\,\mathrm{tr}\,\partial\bar{\partial}(f + \bar{f}) = 0$ where the trace of a (1,1)-form means its $\{F\}$-component; on a Kähler manifold $i\,\mathrm{tr}\,\partial\bar{\partial}$ is the usual Laplacian operator. Now $\sigma + \bar{\sigma} = Dg$ for some function g on U' iff $f + \bar{f}$ is the real part of a holomorphic function. The failure of the Poincaré lemma can then be interpreted as the existence of harmonic functions which are not the real parts of holomorphic functions.

Properties of a domain in the complex manifold Z are related to the nature of its boundary. For U in M, the boundary $N = b(\pi^{-1}U) = \pi^{-1}bU$ is a real hypersurface in Z with an induced CR structure. Take a real function r on M such that $r = 0$, $dr \neq 0$ on bU, so that N is defined by $\pi^\star r = 0$. The Levi form L of N is the restriction of the (1,1)-form $\partial\bar{\partial}(\pi^\star r)$ to the maximal complex subbundle $\Pi = T^{1,0}Z \cap (TN)_c$ in N.

<u>Lemma 11.4</u> If U is strictly convex in the sense that $(\nabla dr)(X,X) > 0$ for all $X \in T(bU)$, then L has non-zero eigenvalues of opposite signs.

<u>Proof</u>. Fix $z \in N$ and take a local basis $\{u^0, u^1\}$ of $\underline{V}_+$ as in (10.3) with $z = \nu(u) \in \pi^{-1}(x)$. Writing $dr = \Sigma\, u^i \otimes v_i$, $v_i \in \Gamma(\underline{V}_-)$,

$$\nabla dr = u^i \otimes \nabla v_i = (u^i \otimes v_j) \otimes \sigma^j_i$$

say. Meanwhile on Z, omitting π^*,

$$\bar{\partial} r = (1 + |\lambda|^2)^{-1}(-u^0 + \bar{\lambda} u^1) \otimes (-v_0 + \lambda v_1)$$

and

$$\partial\bar{\partial} r|_z = -d\bar{\lambda} \wedge (u^1 \otimes v_0) - d\lambda \wedge (u^0 \otimes v_1) + (u^0 \otimes v_j) \wedge \sigma^j_0.$$

If $i : N \hookrightarrow Z$ is the inclusion, $i^* d\lambda$, $i^*(u^1 \otimes v_0)$ form a basis of Π_z, and the lemma follows from the convexity assumption. ■

The above hypothesis is satisfied for example when U is a ball with respect to the distance function of any metric in the conformal class of M. The ensuing property of the Levi form is expressed by saying that $\pi^{-1}U$ is <u>strongly 1-pseudoconvex</u>. Combining results of Andreotti et al. [AG, AN] with theorem 11.3 yields

<u>Theorem 11.5</u> Let M be an anti-self-dual 4-manifold. For suitably convex neighbourhoods U, $H^i(A)$ is infinite-dimensional for $i = 1$, and zero for $i = 2$.

The diagram (11.3) can be "twisted" or tensored with a power ζ^{-r} of the holomorphic line bundle ζ. If the open set U is Spin, r can be any integer; if not r must be even. In other words we now take

$$B^i = \Gamma(\pi^{-1}U, \Lambda^i N \otimes \zeta^{-r}),$$

$$C^i = \Gamma(\pi^{-1}U, \Lambda^{0,i} Z \otimes \zeta^{-r})$$

$$D^i = \Gamma(\pi^{-1}U, \Lambda^{i-1} N \otimes \zeta^{-r-2}).$$

Lemmas 11.1 and 11.2 still apply, and $(A^\star, D)$, $(E^\star, D)$ are complexes of differential operators on U. We enumerate the various cases:

<u>$r = -1$</u> $E^1 = E^2 = E^3 = A^0 = 0$ and $(A^\star, D)$ gives the <u>Dirac operator</u>

$$0 \longrightarrow \underline{V}_- \xrightarrow{D} \underline{V}_+ \longrightarrow 0.$$

This can be defined directly as the composition of covariant differentiation $\underline{V}_- \longrightarrow \underline{V}_- \otimes \underline{T}^\star$ with the symbol homomorphism $V_- \otimes (V_+ \otimes V_-) \longrightarrow \Lambda^2 V_- \otimes V_+ \cong V_+$. In flat space $\mathbb{R}^4 \cong \mathbb{H}$, D is the quaternionic Cauchy-Riemann operator

$$D = \frac{\partial}{\partial x^1} + i\frac{\partial}{\partial x^2} + j\frac{\partial}{\partial x^3} + k\frac{\partial}{\partial x^4}$$

acting on quaternionic-valued functions. Theorem 11.5 then guarantees the local existence of lots of "quaternionic holomorphic" functions.

<u>$r \geq 0$</u> $E^1 = E^2 = E^3 = 0$, and $(A^\star, D)$ becomes

(11.4) $$0 \longrightarrow S^r\underline{V}_+ \xrightarrow{D_0} S^{r+1}\underline{V}_+ \otimes \underline{V}_- \xrightarrow{D_1} S^{r+2}\underline{V}_+ \longrightarrow 0,$$

which is a rearrangement of the Dirac operator

$$0 \longrightarrow \underline{V}_- \otimes S^{r+1}\underline{V}_+ \longrightarrow \underline{V}_+ \otimes S^{r+1}\underline{V}_+ \longrightarrow 0$$

with coefficients in the vector bundle $S^{r+1}\underline{V}_+$. In contrast to H^1 and H^2, $H^0(A) \cong \ker D_0$ generally has finite non-zero dimension. For example for any connected $U \subset \mathbb{R}^4$,

$$\ker D_0 \cong H^0(\mathbb{C}P^3, \zeta^{-r}) \cong S^r(\mathbb{C}^4).$$

Furthermore if U is Einstein with scalar curvature $t \neq 0$ and $r = 2$, $\ker D_0$ is isomorphic to the space of infinitesimal isometries on U. For an infinitesimal isometry can be regarded as a 1-form α such that $\nabla\alpha$ is wholly skew, and the mapping

$$\alpha \longmapsto (\nabla\alpha)_+ \in \Gamma(\underline{\Lambda}^2_+) \cong \Gamma(S^2\underline{V}_+)$$

induces the required isomorphism [S; lemma 6.4].

When $U = M$ is Einstein with $t > 0$ and compact, we know from the last section that the canonical bundle $\kappa \cong \zeta^4$ of Z is negative. Kodaira's vanishing theorem and Serre duality give $H^i(Z, \zeta^{-r}) = 0$ for $i = 1,2,3$, $r = 0,2$, whereas the corresponding vanishing of $H^i(A)$ can be deduced directly using Bochner-type arguments on M. Applying the Atiyah-Singer index theorem to (11.4) for $r = 0,2$ then gives respectively

$$1 = \chi(Z, \mathcal{O}) = \frac{1}{2}(\chi + \tau)$$

$$I = \dim \ker D_0 = \chi(Z, \zeta^{-2}) = 5\chi + 7\tau,$$

where χ, τ are the Euler characteristic and signature of M. These equations are used to establish (10.6).

<u>$r = -2$</u> $E^1 \cong C^\infty(U)$, $E^2 = E^3 = A^0 = A^1 = 0$, $A^2 = \Gamma(U, \Lambda^2 \underline{V}_-) \cong C^\infty(U)$. The coboundary homomorphism

$$(11.5) \qquad 0 \longrightarrow C^\infty(U) \xrightarrow{d} C^\infty(U) \longrightarrow 0$$

turns out to be a certain second order wave operator $\nabla^\star \nabla + \frac{1}{6}t$ [H_3], which has no solutions when $U = M$ is compact with scalar curvature $t > 0$. At the same time the complex (11.4) has cohomology groups $H^i(\pi^{-1}U, \zeta^2)$, $i = 1,2$. The last two facts were extremely relevant in the classification of self-dual vector bundles over S^4 [ADHM].

<u>$r \leqslant -3$</u> $A^0 = A^1 = A^2 = 0$, and $(E^\star, D)$ becomes

$$0 \longrightarrow S^{p+2}\underline{V}_+ \longrightarrow S^{p+1}\underline{V}_+ \otimes \underline{V}_- \longrightarrow S^p \underline{V}_+ \longrightarrow 0$$

where $p = -r-4$. The situations for equal values of p and r are dual with the roles of $A^\star$ and $E^\star$ reversed, whereas $p = r = -2$ is a special case in the middle. The absence of E^0 corresponds to the fact that the line bundle ζ^{-r} has no holomorphic sections (even over a fibre) for $r \leqslant -1$.

The differential operators on an anti-self-dual 4-manifold that we have just defined give rise to certain field equations in physics, and the correspondence between these and sheaf cohomology groups, described also in [EPW, H_3], is one of the corner stones of twistor theory. The author is grateful to M. G. Eastwood for his suggestion of the use of lemma 11.1, which allowed us to treat the left-handed case ($r > -2$) and the right-handed one ($r < -2$) on an equal footing. Our technique of using the Dolbeault complex on Z to induce operators on M also works for the generalized twistor spaces discussed at the end of section 8. In particular a quaternionic Kähler manifold M possesses an elliptic complex of differential operators for each integer r which is merely a lengthened version of the one listed above. For instance taking $r = 3$ produces a resolution of the sheaf of germs of solutions of the quaternionic Cauchy-Riemann equations on M.

12. CONFORMAL STRUCTURE

Many of the properties of a Riemannian 4-manifold M we have examined arise from the splitting

$$\Lambda^2 T^* = \Lambda^2_+ \oplus \Lambda^2_- \tag{12.1}$$

and the subsequent notion of self-duality. Now (12.1) is really a decomposition of representations of the group $SO(4)$, but it is also invariant under the action of the product $CO(4) = \mathbb{R}^+ \times SO(4) \subset GL(4,\mathbb{R})$, where $\mathbb{R}^+$ denotes positive multiples of the identity. If P is the principal bundle of oriented orthonormal frames of M, the enlarged bundle P_i, where $i : SO(4) \hookrightarrow CO(4)$ is the inclusion, consists of frames whose elements are orthogonal vectors with equal but unspecified norms. Moreover these frames are still oriented because we have defined $CO(4)$ to be connected. In other words P_i determines precisely the oriented conformal structure of M.

Vector bundles naturally defined by an oriented conformal structure in 4 dimensions are therefore those associated to P_i by means of a representation of $CO(4)$. Having already tackled $SO(4)$, it remains only to understand the action of $\mathbb{R}^+$. Given any $r \in \mathbb{R}$, let L^r denote the representation of $\mathbb{R}^+$ on the vector space $\mathbb{R}$ in which an element $t \in \mathbb{R}^+$ acts by multiplication by $t^r \in \mathbb{R}^+$. Then a typical $CO(4)$-module has the form

$$S^{p,q;r} = S^p V_+ \otimes S^q V_- \otimes L^r, \quad p + q \text{ even},$$

and r is called its <u>conformal weight</u>. For example the tangent and cotangent bundles can no longer be identified since $\mathbb{R}^+$ acts non-trivially on them; indeed

$$\underline{T} \cong \underline{S}^{1,1;1}, \quad \underline{T}^* \cong S^{1,1;-1}.$$

Following up on proposition 9.5, we shall see next that the twistor space Z of an anti-self-dual Riemannian 4-manifold depends only on the underlying conformal structure. First observe that as a real manifold,

$$Z = \mathbb{P}(\underline{V}_+) = P \times_{SO(4)} \mathbb{P}(V_+) = P_i \times_{CO(4)} \mathbb{P}(V_+)$$

is completely determined by the conformal structure, so it remains to demonstrate

<u>Theorem 12.1</u> Let M be a 4-manifold with an oriented conformal structure. Then any two Riemannian metrics within the conformal class determine <u>identical</u> almost complex structures on Z.

<u>Proof</u>. Let $\nabla, \tilde{\nabla}$ be the Riemannian connections associated to Riemannian metrics g, $\tilde{g} = \lambda^2 g$ on M, where λ is a positive scalar function. Fix a local basis of 1-forms $\{e^1, e^2, e^3, e^4\}$ orthonormal relative to g and suppose that

$$\nabla e^i = e^j \otimes \sigma^i_j, \quad \tilde{\nabla} e^i = e^j \otimes \tilde{\sigma}^i_j.$$

Since $\nabla, \tilde{\nabla}$ are torsion-free,

$$de^i = e^j \wedge \sigma^i_j = e^j \wedge \tilde{\sigma}^i_j,$$

and the components a^i_{jk} of the difference tensor

$$\sigma^i_j - \tilde{\sigma}^i_j = a^i_{jk} e^k$$

are symmetric in j,k. Since $\{\lambda e^i\}$ is orthonormal relative to $\tilde{g}$, the connection forms on the right hand side of

$$\tilde{\nabla}(\lambda e^i) = \lambda e^j \otimes (\tilde{\sigma}^i_j + \delta^i_j \lambda^{-1} d\lambda)$$

are skew, as are σ^i_j. Putting $\lambda^{-1} d\lambda = a_k e^k$ we have

$$a^i_{jk} + a^j_{ik} = 2\delta^i_j a_k$$

which means that in the indices i,j, a^i_{jk} belongs to the Lie algebra

$co(4) \cong \mathbb{R} \oplus so(4)$, and interchanging the indices many times gives

$$a^i_{jk} = \delta^i_j a_k + \delta^i_k a_j - \delta_{jk} a_i .$$

The assignment $a_i \longmapsto a^i_{jk}$ defines a monomorphism

$$T^* \hookrightarrow T \otimes S^2 T^*$$

of $CO(4)$-modules whose image equals

$$co(n)^{(1)} = (T \otimes S^2 T^*) \cap (co(4) \otimes T^*)$$

and is known as the first prolongation of $co(n)$. Thus $\nabla, \tilde{\nabla}$ differ by a section of the associated bundle:

$$(12.2) \qquad \xi = \nabla - \tilde{\nabla} = a^i_{jk} e_i \otimes e^j \otimes e^k \in \Gamma(\underline{co(n)^{(1)}}).$$

Turning attention to theorem 8.1, recall that the almost complex structure on Z is induced from one on the total space of the bundle $\underline{V}_+ \backslash \underline{0}$. However $\mathbb{P}(\underline{V}_+)$ is naturally isomorphic to $\mathbb{P}(\underline{V}_+ \otimes \underline{L}^r)$, so $\underline{V}_+$ may be replaced by $\underline{V}_+ \otimes \underline{L}^r$ for any $r \in \mathbb{R}$. Accordingly let $\{u^0, u^1\}$ be a local basis of $\underline{V}_+ \otimes \underline{L}^r$ with corresponding forms $\omega^i_j, \omega_j; \tilde{\omega}^i_j, \tilde{\omega}_j$ relative to $\nabla, \tilde{\nabla}$ as in the proof of theorem 8.1. By (8.3), ∇ and $\tilde{\nabla}$ define the same almost complex structure on $\underline{V}_+ \otimes \underline{L}^r \backslash \underline{0}$ iff

$$\omega_j - \tilde{\omega}_j = \lambda_i (\omega^i_j - \tilde{\omega}^i_j)$$

belongs to $u \otimes \pi^* \underline{V}_-$ at the point $u = \lambda_k u^k$. This is the case iff

$$\omega^i_j - \tilde{\omega}^i_j = u^i \otimes v_j$$

for some $v_j \in \underline{V}_-$. But the tensor with components $\omega^i_j - \tilde{\omega}^i_j$ is the image of ξ under the mapping

$$\psi : co(4) \otimes T^* \longrightarrow \mathrm{End}\, V_+ \otimes T^*$$

induced from the representation $CO(4) \longrightarrow \mathrm{Aut}(V_+ \otimes L^r)$. Let ι denote the involution of

$$\mathrm{End}\, V_+ \otimes T^* \cong V_+^* \otimes \overbrace{V_+ \otimes V_+}^{\leftrightarrow} \otimes V_- \otimes L^{-1}$$

obtained by interchanging the indicated factors. Then it remains to show that $\psi(co(n)^{(1)}) = \iota(1 \otimes T^\star)$, where $1 \in \operatorname{End} V_+$ is the identity. By Schur's lemma, $\psi(co(n)^{(1)})$ must lie in the submodule isomorphic to $T^\star \otimes \mathbb{R}^2$ spanned by $1 \otimes T^\star$ and $\iota(1 \otimes T^\star)$. But as r varies, $\psi(co(n)^{(1)})$ assumes all subspaces $T^\star \otimes (x,y)$ except the one corresponding to $1 \otimes T^\star$. Therefore for an appropriate choice of conformal weight r, ∇ and $\tilde{\nabla}$ define the same almost complex structure on $\underline{V}_+ \otimes \underline{L}^r \setminus \underline{0}$. The theorem follows. ∎

<u>Corollary</u> The condition $W_+ \equiv 0$ is conformally invariant, and in this case Z is a complex manifold.

Of course it is known that the full Weyl tensor $W = W_+ + W_-$ is conformally invariant. This follows directly from (12.2) and the fact that if $R, \tilde{R}$ are the curvature tensors associated to $\nabla, \tilde{\nabla}$, then

$$R - \tilde{R} = \nabla_1 \xi + [\xi, \xi],$$

where ∇_1 is an extended covariant derivative [AHS]. As a $CO(4)$-module, $co(n)^{(1)} \cong T^\star$ so $R - \tilde{R}$ lies only in submodules of $T^\star \otimes T^\star$. From corollary 5.2, $W = \tilde{W}$ and changing the conformal class of the metric affects only the Ricci tensor. Since the differential operators of the previous section arose from the holomorphic structure of Z, these too can be made conformally invariant by choosing appropriate weights [F].

To conclude we describe natural conformal structures on $\mathbb{HP}^1$ and S^4 without reference to their previously defined Riemannian metrics. The double covering $Sp(1) \times Sp(1) \longrightarrow SO(4)$ determines a homomorphism

$$\mathbb{H}^\star \times \mathbb{H}^\star \longrightarrow CO(4)$$

where $\mathbb{H}^\star$ denotes the group of non-zero quaternions. Indeed by the proof of proposition 1.1, the linear transformation $\phi : \mathbb{H} \longrightarrow \mathbb{H}$ given by $\phi(q) = Aq\bar{B}$ defines an element of $CO(4)$ for any $A, B \in \mathbb{H}^\star$. In terms of the usual homogeneous coordinates $[q_0, q_1]$ of $\mathbb{HP}^1$, let

$q = q_1 q_0^{-1}$ be the function defined on the open set $\mathbb{H}P^1 \setminus [0,1]$. The group $GL(2,\mathbb{H})$ acts on $\mathbb{H}^2$ on the left:

$$q_0 \longmapsto aq_0 + bq_1$$

$$q_1 \longmapsto cq_0 + dq_1,$$

and its centreless version $PGL(2,\mathbb{H})$ acts as projective transformations on $\mathbb{H}P^1$, sending the coordinate q to

$$q' = (cq_0 + dq_1)(aq_0 + bq_1)^{-1}.$$

Then at each point where q, q' are defined,

$$dq' = A\, dq\, B, \quad A, B \in \mathbb{H}^*.$$

This proves that $\mathbb{H}P^1$ admits a conformal structure preserved by the action of $PGL(2,\mathbb{H})$. The corresponding principal $CO(4)$-bundle consists of bases formed by the real components of the quaternionic 1-forms dq'.

To relate the above to S^4 we need

Proposition 12.2 $PGL(2,\mathbb{H}) \cong SO_0(5,1)$

Proof. This is similar to that of proposition 2.3 which deals with the respective maximal compact subgroups. $PGL(2,\mathbb{H})$ is double-covered by $SL(2,\mathbb{H}) = GL(2,\mathbb{H}) \cap SL(4,\mathbb{C})$ which acts on $U = \mathbb{C}^4$. The quaternionic structure map of U makes $\Lambda^2 U$ into a real vector space; indeed if $\{u^1, u^2 = ju^1, u^3, u^4 = ju^3\}$ is a basis of U, then

$$\begin{aligned} \eta^0 &= u^1 \wedge u^2 + u^3 \wedge u^4 & \eta^3 &= u^1 \wedge u^2 - u^3 \wedge u^4 \\ \eta^1 &= i(u^1 \wedge u^3 + u^4 \wedge u^2) & \eta^4 &= u^1 \wedge u^3 - u^4 \wedge u^2 \\ \eta^2 &= i(u^1 \wedge u^4 + u^2 \wedge u^3) & \eta^5 &= u^1 \wedge u^4 - u^2 \wedge u^3 \end{aligned} \tag{12.3}$$

constitute a real basis of $\Lambda^2 U$. Trivializing $\Lambda^4 U$ with $u^1 \wedge u^2 \wedge u^3 \wedge u^4$,

$$h(\eta^i,\eta^j) = \eta^i \wedge \eta^j = \begin{cases} 0 & i \neq j \\ +1 & i = j = 0 \\ -1 & \text{otherwise} \end{cases}$$

defines a real metric of signature (5,1). The action of $SL(2,\mathbb{H})$ on $\Lambda^2 U$ then determines the requires isomorphism, $SO_0(5,1)$ being the connected component of the identity of the group preserving the metric h. ∎

The geometrical significance of the space $\Lambda^2 U$ is that if $C_{5,1}$ denotes the cone

$$\{\gamma \in \Lambda^2 U : \gamma \wedge \gamma = 0, \gamma \text{ real}\} = \{x_i \eta^i : x_0{}^2 = \sum_{i=1}^{5} x_i{}^2\}$$

then the quotient $\hat{C}_{5,1} = {C_{5,1} \backslash 0}/{\mathbb{R}^*}$ can be identified conformally with both $\mathbb{H}P^1$ and S^4. For any $\gamma \in C_{5,1} \backslash 0$ determines an indecomposable element of $\Lambda^2 U$ whose associated complex 2-plane in U is a quaternionic line, i.e. belongs to $\mathbb{H}P^1$. On the other hand h induces a positive definite metric on $\mathbb{R}^5 = \{x_0 = c > 0\}$ and on the sphere $C^{5,1} \cap \mathbb{R}^5$ which gives rise to a conformal structure on $\hat{C}_{5,1}$, independent of c and preserved by $SO_0(5,1)$. In the homogeneous descriptions

$$\mathbb{H}P^1 = \frac{PGL(2,\mathbb{H})}{G_0} \cong \frac{SO_0(5,1)}{G_0} = S^4$$

the isotropy subgroup G_0 is a semidirect product with Lie algebra $co(n) \oplus co(n)^{(1)}$. Unlike in the Riemannian case, points of the corresponding principal G_0-bundle contain second order information and are 2-jets of transformations on the base space.

Ignoring the real structure of U in the proof of proposition 12.2 shows that there is a double covering

$$SL(4,\mathbb{C}) \longrightarrow SO(6,\mathbb{C})$$

of which $SL(2,\mathbb{H}) \longrightarrow SO_0(5,1)$ is a real form. Other real forms include

$$SU(4) \longrightarrow SO(6)$$

$$SL(4,\mathbb{R}) \longrightarrow SO_0(3,3)$$

$$SU(2,2) \longrightarrow SO_0(4,2),$$

corresponding to the other possible signatures on the real vector space $\Lambda^2 U$. The first gives $SU(4) \cong Spin(6)$, completing our list of isomorphisms for the low dimensional Spin groups. Taking maximal compact subgroups of the second gives the isomorphism

$$SO(4) \longrightarrow SO(3) \times SO(3)$$

defined by (2.2). Then we have

$$\frac{GL^+(4,\mathbb{R})}{CO(4)} \cong \frac{SL(4,\mathbb{R})}{SO(4)} \cong \frac{SO_0(3,3)}{SO(3) \times SO(3)}.$$

The first of these homogeneous spaces represents the set of all oriented conformal structures on $\mathbb{R}^4$, whereas the last can be identified with the open set of the Grassmannian of real 3-planes in $\mathbb{R}^6$ on which a metric of signature (3,3) is positive definite. A given conformal structure is mapped to the subspace $\Lambda^2_+ \subset \mathbb{R}^6$, establishing an earlier claim that the conformal structure is determined by the space Λ^2_+.

The homogeneous spaces $\hat{C}_{3,3}, \hat{C}_{4,2}, \hat{C}_{5,1}$ of rays in the respective cones are real forms of the complex Grassmannian of 2-planes in $\mathbb{C}^4$ which parametrizes projective lines in the twistor space $\mathbb{C}P^3$. Indeed $\hat{C}_{3,3} \approx S^2 \times S^2$ is the real Grassmannian, whereas $\hat{C}_{4,2} \approx S^3 \times S^1$ is a conformal compactification $\bar{M}$ of Minkowski space M [P, V, We].

Writing $C_{4,2} = \{x_0^2 + x_5^2 = \sum_{i=1}^{4} x_i^2\}$, M can be identified with the slice

$$C_{4,2} \cap \{x_0 = c > 0\} \hookrightarrow \hat{C}_{4,2},$$

the Lorentzian metric being induced from the metric of type (4,2) on $\Lambda^2 U$. Points of $\bar{M}$ parametrize those complex lines that lie in the real 5-dimensional hypersurface N of $\mathbb{C}P^3$ defined by the vanishing of the form on $\mathbb{C}^4$ preserved by SU(2,2). Conversely, points of N parametrize null lines in $\bar{M}$; these represent paths of zero rest-mass particles, and obviously depend only upon the conformal structure. This correspondence was developed by Penrose, and provided the original motivation for introducing the twistor space.

REFERENCES

[A] Adams, J.F.: Lectures on Lie Groups. W.A. Benjamin, New York, 1969.

[AG] Andreotti, A., Grauert, H.: Théorèmes de finitude pour la cohomologie des espaces complexes. Bull. Soc. Math. Fr., 90 (1962) 193-259.

[AN] Andreotti, A., Norguet, F.: Problème de Levi et convexité holomorphe pour les classes de cohomologie. Ann. Sc. Norm. Super. Pisa, Cl. Sci., IV. Ser., 10 (1966) 197-241.

[ADHM] Atiyah, M.F., Drinfeld, V.G., Hitchin, N.J., Manin, Yu.I.: Construction of instantons. Phys. Lett. 65A, (1978) 185-187.

[AHS] Atiyah, M.F., Hitchin, N.J., Singer, I.M.: Self-duality in four-dimensional Riemannian geometry. Proc. R. Soc. Lond., Ser. A, 362 (1978) 425-461.

[AS] Atiyah, M.F., Singer, I.M.: The index of elliptic operators: III. Ann. Math., II. Ser., 87 (1968) 546-604.

[BO] Bérard Bergery, L., Ochiai, T.: On some generalisations of the construction of twistor spaces. Durham-LMS Symposium, 1982.

[B] Blanchard, A.: Sur les variétés analytiques complexes. Ann. Sci. Ec. Norm. Super., IV. Ser., 73 (1956) 157-202.

[BH] Borel, A., Hirzebruch, F.: Characteristic classes and homogeneous spaces II. Am. J. Math., 81 (1959) 315-382.

[C_1] Calabi, E.: On Kähler manifolds with vanishing canonical class, in Algebraic Geometry and Topology, in honor of Lefschetz. Princeton University Press, Princeton, 1957

[C_2] Calabi, E.: Métriques kahlériennes et fibrés holomorphes. Ann. Sci. Ec. Norm. Super., IV. Ser., 12 (1979) 269-294.

[D] Derdzinski, A.: Self-dual Kähler manifolds and Einstein manifolds of dimension four. To appear in Compos. Math.

[EPW] Eastwood, M.G., Penrose, R., Wells, R.O., Jr.: Cohomology and massless fields. Commun. Math. Phys., 78 (1981) 305-351.

[EL] Eells, J., Lemaire, L.: A report on harmonic maps. Bull. Lond. Math. Soc., 10 (1978) 1-68.

[F] Fegan, H.D.: Conformally invariant first order differential operators. Q.J. Math., Oxf. II. Ser., 27 (1976) 371-378.

[FK] Friedrich, T., Kurke, H.: Compact four-dimensional self-dual Einstein manifolds with positive scalar curvature. Math. Nachr., 106 (1982) 271-299.

[G] Goldberg, S.I.: Curvature and Homology. Academic Press, New York-London, 1962.

[Gr] Gray, A.: Some examples of almost Hermitian manifolds. Ill. J. Math., 10 (1966) 353-366.

[GrH] Gray, A., Hervella, L.M.: The sixteen classes of almost Hermitian manifolds and their linear invariants. Ann. Mat. Pura Appl., IV. Ser., 123 (1980) 35-58.

[GH] Griffiths P.A., Harris J.: Principles of Algebraic Geometry. Wiley, New York, 1978.

[H_1] Hitchin, N.J.: Compact four-dimensional Einstein manifolds. J. Diff. Geom. 9 (1974) 435-441.

[H_2] Hitchin, N.J.: Polygons and gravitons. Math. Proc. Camb. Philos. Soc., 85 (1979) 465-476.

[H_3] Hitchin, N.J.: Linear field equations on self-dual spaces. Proc. R. Soc. Lond., Ser. A, 370 (1980) 173-191.

[H_4] Hitchin, N.J.: Kählerian twistor spaces. Proc. Lond. Math. Soc., III. Ser., 43 (1981) 133-150.

[K] Kobayashi, S.: Remarks on complex contact manifolds. Proc. Am. Math. Soc., 10 (1959) 164-167.

[KN] Kobayashi, S., Nomizu, K.: Foundations of Differential Geometry. 2 volumes, Interscience, New York, 1963, 1969.

[KO] Kobayashi, S., Ochiai, T.: Characterizations of complex projective spaces and hyperquadrics. J. Math. Kyoto Univ., 13 (1973) 31-47.

[Ko] Kodaira, K.: A theorem of completeness of characteristic systems for analytic families of compact submanifolds of complex manifolds. Ann. Math., II. Ser., 75 (1962) 146-162.

[LB] LeBrun, C.R.: H-Space with a cosmological constant. Proc. R. Soc. Lond., Ser. A, 380 (1982) 171-185.

[M] Matsushima, Y.: Sur la structure du groupe d'homéomorphismes analytiques d'une certaine variété Kaehlérienne. Nagoya Math. J., 11 (1957) 145-150.

[NN] Newlander A., Nirenberg L.: Complex analytic coordinates in almost complex manifolds. Ann. Math., II. Ser., 65 (1957) 391-404.

[P] Penrose, R.: The twistor programme. Rep. Math. Phys., 12 (1977) 65-76.

[S] Salamon, S.: Quaternionic Kähler manifolds. Invent. Math., 67 (1982) 143-171.

[ST] Singer, I.M., Thorpe, J.A.: The curvature of 4-dimensional Einstein spaces, in Global Analysis, in honor of Kodaira. Princenton University Press, Princeton, 1969.

[So] Sommese, A.: Quaternionic manifolds. Math. Ann., 212 (1975) 191-214.

[St] Struik, D.J.: On the sets of principal directions in a Riemannian manifold of four dimensions. J. Math. Phys. M.I.T. 7 (1927-28) 193-197.

[T] Thurston, W.P.: Some simple examples of symplectic manifolds. Proc. Am. Math. Soc., 55 (1976) 467-468.

[V] Veblen, O.: Geometry of four component spinors. Proc. Natl. Acad. Sci. USA, 19 (1933) 503-517.

[W] Weil, A.: Sur les théorèmes de deRham. Comment. Math. Helv., 26 (1952) 119-145.

[We] Wells, R.O., Jr.: Complex geometry and mathematical physics. Bull. Am. Math. Soc., New Ser., 1 (1979) 296-336.

[Y] Yau, S.T.: On the Ricci curvature of compact Kähler manifolds and complex Monge-Ampère equations I. Commun. Pure Appl. Math., 31 (1978) 339-411.

[Z] Zelobenko, D.P.: Compact Lie Groups and their Representations. Translation of mathematical monographs 40, Am. Math. Soc., Providence, 1973.

JEAN-PIERRE VIGUÉ

DOMAINES BORNÉS SYMÉTRIQUES

PREFACE

Les notes qui vont suivre sont la rédaction d'une série de conférences que j'ai faites à la Scuola Normale Superiore di Pisa durant l'été 1982. J'ai essayé de donner les résultats essentiels, et autant que possible, des démonstrations complètes ou du moins des idées de démonstrations. Volontairement, je me suis limité à étudier les automorphismes des domaines bornés et des domaines bornés symétriques, et je n'ai pas abordé la question des variétés normées symétriques pour lesquelles de nombreux resultats se généralisent (voir [12], [16] et [23]).

Puisque l'occasion m'en est donnée, je suis heureux de remercier M. Vesentini qui m'a invité à la Scuola Normale Superiore et qui a permis la réalisation de ce travail.

CHAPITRE I

Automorphismes analytiques d'un domaine borné d'un espace de Banach complexe: la topologie de la convergence uniforme locale.

1.1 Introduction: le cas de la dimension finie.

L'étude du groupe des automorphismes analytiques d'un domaine borné D de $\mathbb{C}^n$ a été faite par Henri Cartan aux alentours de 1930-1935 (voir [3 à 7]). Soit donc D un domaine borné de $\mathbb{C}^n$. On munit le groupe $G(D)$ de la topologie de la convergence uniforme sur tout compact de D. Alors $G(D)$ est un groupe topologique localement compact. [Plus précisément, si K est un compact de D, et si a est un point de D,

$$\{f \in G(D) \mid f(a) \in K\}$$

est un compact de $G(D)$]. De plus, l'application

$$G(D) \times D \longrightarrow D$$

$$(f,x) \longmapsto f(x)$$

est continue.

Henri Cartan a montré également que le groupe $G(D)$ a, de façon naturelle, une structure de groupe de Lie réel compatible avec sa topologie, et telle que l'application

$$G(D) \times D \longrightarrow D$$

$$(f,x) \longmapsto f(x)$$

soit analytique réelle (voir [6]).

En utilisant les résultats de Henri Cartan, Elie Cartan [2] a donné, en 1935, une classification complète des domaines bornés symétriques de $\mathbb{C}^n$. Rappelons d'abord les résultats suivants: Soit D un domaine borné de $\mathbb{C}^n$, soit a un point de D, et soit s un automorphisme analytique de D. On dit que s est une symétrie par rapport au point a de D, si $s^2 = \mathrm{id}$, et si a est un point invariant isolé de s. On montre qu'un tel s, s'il existe, est forcément unique, et on le note s_a. Un domaine borné D est dit symétrique si, pour tout point a de D, il existe une symétrie s_a par rapport au point a. Elie Cartan a montré alors que, si D est un domaine borné symétrique, D est un domaine homogène [ce qui signifie que le groupe $G(D)$ agit transitivement]. Ainsi, le groupe $G(D)$ est un gros groupe et il contient beaucoup d'informations sur D. Comme il connaissait très bien les groupes de Lie, Elie Cartan montra le résultat suivant: tout domaine borné symétrique est isomorphe à un produit fini de domaines bornés symétriques irréductibles [dire que D est irréductible signifie que D n'est pas isomorphe à un produit $D_1 \times D_2$] et les domaines bornés symétriques irréductibles sont du type suivant:

1) 4 grandes classes (appelées domaines classiques),
2) 2 domaines exceptionnels (correspondant à des groupes de Lie exceptionnels).

Les domaines classiques peuvent se réaliser de la façon suivante (Harris [9]). Soient X et Y deux espaces de Hilbert de dimension finie. Soit $\mathcal{L}(X,Y)$ l'espace des applications linéaires continues de X dans Y, muni de la norme habituelle. Pour tout $x \in \mathcal{L}(X,Y)$, soit $x^\star \in \mathcal{L}(Y,X)$ l'adjoint de x. Soit E un sous-espace de $\mathcal{L}(X,Y)$ tel que, pour tout $x \in E$, $x x^\star x \in E$. Soit B la boule-unité ouverte de $\mathcal{L}(X,Y)$. Alors $B \cap E$ est un domaine borné symétrique de E, et tout domaine borné symétrique classique peut se réaliser de cette façon.

Bien sûr, ces définitions ont encore un sens, même si on ne suppose plus que X et Y sont de dimension finie. Alors, $\mathcal{L}(X,Y)$ est l'espace de Banach des applications linéaires continues de X dans Y, E est un sous-espace vectoriel fermé de $\mathcal{L}(X,Y)$ tel que, pour tout $x \in E$, $x x^\star x \in E$, et Harris [9] montre que $B \cap E$ est un domaine borné symétrique de E.

Donnons un autre exemple en dimension infinie. Soit $\mathcal{C}(K,\mathbb{C})$ l'espace de Banach des fonctions continues à valeurs complexes sur un espace compact K, muni de la norme

$$\|f\| = \sup_{s \in K} |f(s)|$$

La boule-unité ouverte B de $\mathcal{C}(K,\mathbb{C})$ est un domaine borné symétrique. En effet, B est symétrique par rapport à l'origine, et B est homogène. Un automorphisme qui envoie l'origine 0 sur la fonction a est donné par

$$f \longmapsto \frac{f + a}{1 + \bar{a}f} .$$

Ceci suffit à montrer que B est homogène.

Ainsi, l'étude du groupe des automorphismes analytiques d'un domaine borné et d'un domaine borné symétrique a un intérêt aussi dans un espace de Banach complexe, et nous allons le faire dans ces notes. Commençons par définir la topologie de $G(D)$.

1.2. Topologie de la convergence uniforme locale.

Soit D un domaine borné d'un espace de Banach complexe E. Soit $H(D,D)$ l'ensemble des applications holomorphes de D dans D, et soit $G(D)$ le groupe des automorphismes analytiques de D. Commençons par la définition suivante.

<u>Définition 1.2.1</u> Un sous-ensemble A de D est dit complètement intérieur à D (et on note $A \subset\subset D$) si la distance de A à la frontière de D est strictement positive.

Dire qu'une boule $B(a,r) \subset\subset D$ révient à dire qu'il existe $R > r$ tel que $B(a,R)$ soit contenue dans D.

<u>Théorème 1.2.2</u> Soit B une boule non vide $\subset\subset D$. Considérons sur $H(D,D)$ la structure uniforme et la topologie de la convergence uniforme

sur B. Elles ne dépendent pas du choix de la boule $B \subset\subset D$, et nous les appellerons structure uniforme et topologie de la convergence uniforme locale.

Si B_1 et B_2 sont deux boules concentriques $\subset\subset D$, le fait que la topologie (resp. structure uniforme) de la convergence uniforme sur B_1 et B_2 sont égales découle du théorème des 3 cercles d'Hadamard. Le cas général s'en déduit par un argument de connexité.

Bien sûr, si E est de dimension finie, on retrouve sur $H(D,D)$ et sur $G(D)$ la topologie de la convergence uniforme sur tout compact de D.

A l'avenir, nous considérons le groupe $G(D)$ muni de la topologie de la convergence uniforme locale. Nous avons le théorème suivant.

<u>Théorème 1.2.3</u> Le groupe $G(D)$, muni de la topologie de la convergence uniforme locale, est un groupe topologique. L'application

$$G(D) \times D \longrightarrow D$$

$$(f,x) \longmapsto f(x)$$

est continue. Muni de la structure uniforme gauche (resp. droite) associé, $G(D)$ est complet.

Le fait que $G(D)$ est complet se deduit du résultat plus précis suivant

<u>Théorème 1.2.4</u> Soit $(f_n)_{n \in \mathbb{N}}$ une suite de Cauchy pour la structure uniforme de la convergence uniforme locale. Supposons qu'il existe $a \in D$ tel que $f_n(a)$ converge vers un point b de D. Alors f_n converge vers un automorphisme analytique f de D.

L'essentiel de la démonstration [18] consiste à montrer que la suite des (f_n^{-1}) est aussi une suite de Cauchy.

Nous avons le théorème d'unicité suivant.

<u>Théorème 1.2.5</u> (H. Cartan). Soit $a \in D$. Soit $f \in H(D,D)$ telle que

$f(a) = a$, $f'(a) = \mathrm{id}$. Alors f est l'application identique.

Démonstration. Soit

$$f(a + x) = a + x + \sum_{p \geq 2} P_p(x)$$

le développement en séries de f ou voisinage de a. Si $f \neq \mathrm{id}$, il existe un plus petit entier $k \geq 2$ tel que $P_k \not\equiv 0$. Soit $f^n = f\circ\ldots\circ f$ la $n^{\underline{ie}}$ itérée de f. Son développement en série au voisinage de a vaut

$$f^n(a + x) = a + x + n\, P_k(x) + \ldots,$$

et d'après les inégalités de Chauchy, $\|n\, P_k\| \leq M$, ce qui prouve que $P_k \equiv 0$, et le théorème est démontré.

Ce théorème montre déja que l'application

$$G(D) \xrightarrow{\varphi_a} D \times \mathcal{L}(E,E)$$

$$f \longmapsto (f(a), f'(a))$$

est injective. Cette application φ_a caractérise aussi la topologie de $G(D)$ comme le prouve le théorème suivant.

Théorème 1.2.6 L'application φ_a est un homéomorphisme de $G(D)$ sur son image.

Ce théorème est une conséquence du résultat technique suivant [18].

Lemme 1.2.7 Soit D un domaine borné. Soit $a \in D$, et soit A une partie $\subset\subset D$. Soit B une boule $\subset\subset D$. Alors il existe une constante K telle que, pour tout $f \in G(D)$, pour tout $g \in G(D)$ tels que $f(a)$ et $g(a)$ appartiennent à A, on ait

$$\|f - g\|_B \leq K \sup(\|f(a) - g(a)\|, \|f'(a) - g'(a)\|)$$

Pour conclure ce chapitre, nous allons traiter un exercice.

<u>Exercice 1.2.8</u> On dit qu'on domaine D est cerclé si l'origine 0 appartient à D et si, pour tout $x \in D$, pour tout $\lambda \in \mathbb{C}$, $|\lambda| = 1$, $\lambda x \in D$. Nous avons alors le résultat suivant:

Soient D_1 et D_2 deux domaines cerclés bornés. Soit $f : D_1 \longrightarrow D_2$ un isomorphisme analytique de D_1 sur D_2 tel que $f(0) = 0$. Alors f est linéaire.

<u>Démonstration</u>. Pour tout $\theta \in \mathbb{R}$, considerons

$$\varphi(x) = e^{-i\theta} f^{-1}(e^{i\theta} f(x))$$

Il est clair que φ envoie D_1 dans D_1 et que $\varphi(0) = 0$, $\varphi'(0) = \mathrm{id}$. Ainsi $\varphi(x) = x$, ce que entraîne que

$$f(e^{i\theta} x) = e^{i\theta} f(x).$$

Il suffit alors de considérer le developpement de f au voisinage de l'origine pour voir que f est linéaire.

CHAPITRE II

L'algébre de Lie des transformations infinitésimales d'un domaine borné D.

Nous avons défini au chapitre précédent une topologie sur le groupe $G(D)$ des automorphismes analytiques d'un domaine borné D d'un espace de Banach complexe E. La question qui se pose maintenant est de savoir si $G(D)$ est un groupe de Lie. Plus précisément, la question est la suivante: Existe - t - il une structure de groupe de Lie réel sur $G(D)$ telle que la topologie sous-jacente soit la topologie de la convergence uniforme locale et telle que l'application

$$G(D) \times D \longrightarrow D$$

$$(f,x) \longmapsto f(x)$$

soit analytique réelle?

Nous verrons, à la fin de ce chapitre, que la réponse à cette question n'est pas toujours oui. Nous allons commencer par étudier une question un peu différente. Soit D un domaine borné d'un espace de Banach complexe E. Soit Γ un groupe de Lie réel connexe. Je dirai que Γ agit sur D par automorphismes analytiques s'il existe un homorphisme de groupe continu $\rho_\Gamma : \Gamma \longrightarrow G(D)$ telle que l'application induite

$$\Gamma \times D \longrightarrow D$$

$$(g,x) \longmapsto g\,x$$

soit analytique-réelle.

Les (Γ,ρ_Γ) sont les objets d'une catégorie dont les morphismes $(\Gamma_1,\rho_{\Gamma_1}) \longrightarrow (\Gamma_2,\rho_{\Gamma_2})$ sont les morphismes analytiques $\Gamma_1 \xrightarrow{f} \Gamma_2$

tels que le diagramme

$$\begin{array}{ccc} \Gamma_1 & \xrightarrow{\ f\ } & \Gamma_2 \\ & \rho_{\Gamma_1} \searrow \quad \swarrow \rho_{\Gamma_2} & \\ & G(D) & \end{array}$$

soit commutatif. Un objet final de cette catégorie s'appelera un plus grand groupe de Lie connexe agissant sur D par automorphismes analytiques.

Nous allons montrer le résultat suivant ([18] et [16]).

<u>Théorème 2.1.</u> Il existe un plus grand groupe de Lie connexe agissant sur D par automorphismes analytiques.

Le théorème 2.1 se démontre de la façon suivante. Considérons un groupe à un paramètre réel d'automorphismes analytiques de D, c'est-à-dire un homomorphisme continu

$$\mathbb{R} \xrightarrow{\varphi} G(D)$$

tel que $(t,x) \longmapsto \varphi(t)\cdot x = f(t,x)$ soit analytique-réel. A un tel groupe, on associe une transformation infinitésimale ψ qui est définie de la façon suivante

$$\psi(x) = \frac{\partial}{\partial t} f(0,x) = \lim_{t \to 0} \frac{f(t,x)-x}{t} .$$

Rappelons que, étant donné la transformation infinitésimale ψ, on retrouve le groupe a un paramètre associé de la façon suivante:

$$(t,x) \longmapsto f(t,x)$$

est la solution de l'équation différentielle

$$\frac{dx}{dt} = \psi(x)$$

prenant en $t = 0$ la valeur x.

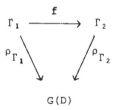

Soit $\mathfrak{g}(D) \subset H(D,E)$ l'ensemble des champs de vecteurs associés a tous les groupes à un paramètre réel d'automorphismes analytiques de D. Le résultat principal est le suivant.

<u>Théorème 2.2.</u> $\mathfrak{g}(D)$ est une algèbre de Lie banachique réelle.

Ce résultat signifie que $\mathfrak{g}(D)$ est un sous-espace vectoriel réel de $H(D,E)$, que l'on peut définir une norme sur E qui en fait un espace de Banach [Il suffit de prendre la norme de la convergence uniforme sur une boule B non vide $\subset\subset D$, et deux telles normes sont équivalentes], et enfin que $\mathfrak{g}(D)$ est fermé sous l'action du crochet. Les arguments essentiels dans cette démonstration sont les deux propositions suivantes.

<u>Proposition 2.3.</u> Soit a un point de D. L'application θ_a:

$$\mathfrak{g}(D) \longrightarrow D \times \mathcal{L}(E,E)$$

$$\psi \longmapsto (\psi(a), \psi'(a))$$

est injective, et est un isomorphisme d'espaces de Banach réels de $\mathfrak{g}(D)$ sur son image.

La proposition 2.3 nous donne un théorème d'unicité pour l'algèbre de Lie $\mathfrak{g}(D)$. Nous avons aussi besoin d'un théorème qui nous permet de construire des éléments de l'algèbre de Lie $\mathfrak{g}(D)$ à partir d'éléments du groupe $G(D)$.

<u>Théorème 2.4.</u> Soit (f_k) une suite d'éléments de $G(D)$ convergeant vers l'identité. Soit a un point de D, et soit,

$$\psi_k = 2^k(f_k - \mathrm{id}).$$

Si $\psi_k(a) \longrightarrow b \in E$ et si $\psi_k'(a) \longrightarrow g \in \mathcal{L}(E,E)$, alors il existe $\psi \in \mathfrak{g}(D)$ tel que $\psi(a) = b$ et $\psi'(a) = g$. De plus, ψ_k converge vers ψ uniformément sur toute boule $B \subset\subset D$.

Il est facile alors de montrer que le groupe Γ_D engendré par les groupes à un paramètre d'automorphismes de D peut être muni d'une structure de groupe de Lie réel connexe d'algèbre de Lie $\mathfrak{g}(D)$. Le groupe Γ_D est le plus grand groupe de Lie connexe agissant sur D par automorphismes analytiques.

L'application $\Gamma_D \hookrightarrow G(D)$ est continue, et $G(D)$ a une structure de groupe de Lie dans laquelle les voisinages de l'identité sont les images des voisinages de l'identité dans Γ_D. Nous la noterons $G_{an}(D)$.

Il est clair que $G(D)$ a une structure de groupe de Lie réel compatible avec sa topologie si et seulement si l'application $\Gamma_D \hookrightarrow G(D)$ ouverte.

Rappelons le

<u>Théorème 2.5.</u> (H. Cartan [6]). Soit D un domaine borné de $\mathbb{C}^n$. Alors $G(D)$ a une structure de groupe de Lie compatible avec sa topologie.

<u>Idée de la démonstration</u>. D'après ce que nous venons de dire, il suffit en fait de montrer que, pour tout voisinage U de l'identité dans Γ_D, l'image $\rho(U)$ de U dans $G(D)$ contient un voisinage de l'identité. Faisons la démonstration par l'absurde. Supposons qu'il existe une suite (f_k) d'automorphismes de D convergeant vers l'identité et n'appartenant pas à $\rho(U)$. Munissons $G(D)$ d'une distance d invariante par translation à gauche et qui définit sa topologie. Quitte à multiplier f_k a gauche par un g_k convenablement choisi, on trouve une suite h_k convergeant vers l'identité et une suite $\varepsilon_k \longrightarrow 0$ telles que

$$d(h_k,id) < d(h_k,\rho(U))\ (1+\varepsilon_k). \tag{1}$$

Soit a un point de D. Quitte à extraire une sous-suite de la suite h_k, on peut supposer qu'il existe une suite d'entiers m_k tels que, si

$$\psi_k = 2^{m_k}(h_k - id),$$

la suite $(\psi_k(a),\psi_k'(a))$ soit une suite bornée de $\mathbb{C}^n \times \mathcal{L}(\mathbb{C}^n,\mathbb{C}^n)$ non adhérente à 0. Par compacité, on peut en extraire une sous-suite

$(\psi_{k_i}(a), \psi'_{k_i}(a))$ convergente. D'après le théorème 2.4, il lui est associé $\psi \in g(D)$. Considérons le groupe à un paramètre $f_\psi(t,\cdot)$ associé à ψ, qui est contenu dans $\rho(U)$ pour t assez petit. On verifie alors que, pour tout i assez grand, il existe t_i tel que $d(h_{m_i}, f_\psi(t_i,\cdot))$ soit très petit devant $d(h_{m_i}, id)$ ce qui contredit l'inégalité (1). Le théorème est démontré.

Le résultat de Henri Cartan ne se généralise pas à la dimension infinie, comme le montre l'exemple suivant.

Exemple. Soit $l^\infty(\mathbb{N})$ l'espace de Banach des suites bornées, muni de la norme de la convergence uniforme. Soit B la boule unité ouverte de $l^\infty(\mathbb{N})$. Le résultat suivant se déduit facilement de [7] et [10] (voir aussi le chapitre IV de ce travail).

Théorème 2.6. L'ensemble des transformations

$$(x_n) \longmapsto f((x_n)) = \left(e^{i\theta_n} \frac{x_n + a_n}{1 + \bar{a}_n x_n}\right)_{n \in \mathbb{N}}$$

pour toute suite de nombres réels θ_n et pour tout suite de nombres complexes (a_n) tels que $\|(a_n)\| < 1$ est un voisinage de l'identité dans $G(B)$.

Remarquons que $G(B)$ est un groupe de Lie, mais nous démontrerons plus tard un résultat plus général.

Soit

$$A_n = \{(0,\ldots,0,\frac{1}{2}\, e^{\frac{2ik\pi}{n+2}}, 0,\ldots 0) \mid k = 0\ldots,n+1\}.$$

Soit $A = \bigcup_{n \in \mathbb{N}} A_n$, et soit $D = B - A$. Il est facile de voir que les automorphismes analytiques de D sont la restriction à D des automorphismes de B laissant A fixe. On en déduit que tout automorphisme analytique f de D, suffisamment proche de l'identité, est de la forme suivante:

$$f((x_n)) = \left(e^{\frac{2i\pi k_n}{n+2}} x_n\right)_{n \in \mathbb{N}}, \quad k_n = 0,\ldots,n+1.$$

Nous avons donc montré le

<u>Théorème 2.7.</u> Le groupe $G(D)$ est complètement discontinu non discret. En particulier ce n'est pas un groupe de Lie.

CHAPITRE III

Domaines bornés symétriques.

Les difficultés rencontrées au chapitre précédent proviennent sans doute du fait que le domaine D n'a pas assez d'automorphismes analytiques. Aussi, nous allons maintenant étudier les domaines bornés symétriques. Nous montrerons que, dans ce cas, le groupe $G(D)$ a une structure de groupe de Lie réel compatible avec sa topologie. Nous montrerons aussi que tout domaine borné symétrique est isomorphe à un domaine borné cerclé étoilé.

3.1. Définitions et premières propriétés.

Proposition et définition 3.1.1. Soit D un domaine borné d'un espace de Banach complexe, et soit a un point de D. Soit $s \in G(D)$ un automorphisme analytique de D. On dit que s est une symétrie par rapport au point $a \in D$ si s satisfait à une des trois conditions équivalentes suivantes:

(i) $s^2 = \mathrm{id}$, et a est un point invariant isolé de s;

(ii) $s(a) = a$ et $s'(a) = -\mathrm{id}$;

(iii) il existe une carte locale u d'un voisinage U de a dans D, telle que $u(a) = 0$, et que, dans cette carte, s soit linéaire, égal à $-\mathrm{id}$.

Un tel s, s'il existe, est unique; on l'appelle la symétrie par rapport au point a, et on la note s_a.

Idéé de la démonstration. D'après le théorème 1.2.5, s, s'il existe, est unique. D'autre part, il est clair que (iii) entraîne (i) et (ii). La réciproque se démontre en construisant une carte locale φ dans laquelle s est linéaire. Il suffit de prendre

$$\varphi(x) = \frac{1}{2}\left[(x - a) + s'(a)^{-1}\cdot(s(x) - a)\right].$$

Définition 3.1.2. On dit qu'un domaine borné D est symétrique si, pour tout point a de D, il existe une symétrie $s_a \in G(D)$ par rapport au point a. On dit que D est homogène si le groupe $G(D)$ agit transitivement, c'est-à-dire, si, pour tout couple de points (a,b) de D, il existe $f \in G(D)$ tel que $f(a) = b$.

Il est facile de voir que tout domaine borné homogène, symétrique par rapport à un point a de D, est symétrique. D'autre part, d'après Piatetsky-Chapiro [14], il existe, déjà en dimension finie, des domaines bornés homogènes non symétriques.

Soit D un domaine borné symétrique d'un espace de Banach complexe E. Nous allons montrer que D est homogène. Comme D est connexe, il suffit de montrer que, pour tout $a \in D$, l'orbite de a sous l'action de $G(D)$ est ouverte. Ce résultat est une conséquence du théorème suivant.

Théorème 3.1.3. Soit $G_{an}(D)$ le groupe des automorphismes analytiques de D, muni de sa structure de groupe de Lie réel. Soit a un point de D. Alors, l'application

$$G_{an}(D) \longrightarrow D$$

$$g \longmapsto g(a)$$

est, au voisinage de l'identité, une submersion directe.

Pour montrer ce théorème, il suffit de montrer, d'après le théorème d'inversion locale, que l'application linéaire tangente

$$\mathfrak{g}(D) \longrightarrow E$$

$$\psi \longmapsto \psi(a)$$

est un épimorphisme direct.

Pour cela, nous commençons par montrer la

Proposition 3.1.4. L'application

$$D \longrightarrow G(D)$$

$$a \longmapsto s_a$$

est continue.

Démonstration. On à

$$\|s_b(a) - s_a(a)\| = \|s_b(a) - a\|$$

$$\leq \|s_b(a) - s_b(b)\| + \|b - a\|,$$

et d'après les inégalites de Cauchy, on trouve

$$\|s_b(a) - s_a(a)\| \leq (K_1 + 1)\|b - a\|.$$

De même on montre

$$\|s'_b(a) - s'_a(a)\| = \|s'_b(a) - s'_b(b)\| \leq K_2\|b - a\|.$$

Le théorème se déduit alors du lemme 1.2.7.

Soit D un domaine borné symétrique par rapport à l'origine $0 \in D$. La symétrie s_o agit par automorphisme intérieur sur l'algébre de Lie $\mathfrak{g}(D)$ et définit une décomposition directe

$$\mathfrak{g}(D) = \mathfrak{g}(D)^+ \oplus \mathfrak{g}(D)^-,$$

où

$$\mathfrak{g}(D)^+ = \{\psi \in \mathfrak{g}(D) \,|\, s_o \cdot \psi = \psi\}$$

$$= \{\psi \in \mathfrak{g}(D) \,|\, \psi(0) = 0\},$$

$$\mathfrak{g}(D)^- = \{\psi \in \mathfrak{g}(D) \,|\, s_o \cdot \psi = - \psi\}.$$

De plus, d'après la proposition 2.3, l'application

$$\mathfrak{g}(D)^- \longrightarrow E$$

$$\psi \longmapsto \psi(0)$$

est un isomorphisme d'espaces de Banach réels de $g(D)^-$ sur son image.

Soit maintenant D un domaine borné symétrique, et supposons que l'origine 0 appartient à D. Pour montrer que D est homogène, il suffit de montrer la proposition suivante.

<u>Proposition 3.1.5.</u> L'application

$$g(D)^- \longrightarrow E$$

$$\psi \longmapsto \psi(0)$$

est un isomorphisme d'espaces de Banach réels de $g(D)^-$ sur E.

<u>Idée de la démonstration</u>. Soit $b \in E$. On déduit de la proposition 3.4.1 que $s_{\frac{b}{2^k}} \circ s_o \longrightarrow \mathrm{id}$.

Soit alors

$$\psi_k = 2^k \left(s_{\frac{b}{2^k}} \circ s_o - \mathrm{id} \right).$$

On montre alors que $\psi_k(0) \longrightarrow 2b$, et que, à condition de se placer dans une carte locale où s_o est linéaire, $\psi_k'(0) \longrightarrow 0$. D'après le théorème 2.4, il existe donc $\psi \in g(D)^-$, tel que $\psi(0) = b$. La proposition est démontrée.

Pour tout $b \in E$, nous noterons X_b l'unique élément de $g(D)^-$ tel que $X_b(0) = b$. D'après la terminologie de Elie Cartan, les éléments de $g(D)^-$ s'appellent des transvections infinitésimales, les éléments de $g(D)^+$ sont des rotations infinitésimales.

<u>Exemple 3.1.6.</u> Soient X et Y deux espaces de Hilbert (munis de la norme habituelle), et soit B la boule-unité ouverte de l'espace de Banach $\mathcal{L}(X,Y)$ des applications linéaires continues de X dans Y. D'après Harris [9], B est un demaine borné symétrique. L'application

qui envoie 0 sur $a \in B$ est donné par

$$x \longmapsto f_a(x) = (1 - aa^\star)^{-\frac{1}{2}}(x + a)(1 + a^\star x)^{-1}(1 - a^\star a)^{\frac{1}{2}},$$

où $a^\star$ designe l'adjoint de a. Les transvections infinitésimales associées sont:

$$x \longmapsto X_b(x) = b - xb^\star x.$$

Ainsi, les transvections infinitésimales sont des polynômes de degré 2, le groupe d'isotropie de l'origine $G_o(D)$ est linéaire, et $g(D)^+$ est linéaire. En fait, ce résultat est général, comme nous allons le voir maintenant.

3.2. Isomorphisme d'un domaine borné symétrique sur un domaine cerclé borné.

Il découle du théorème de Liouville que, si $\psi \in g(D)$ est non nul, $i\psi \notin g(D)$. On peut donc considérer le complexifié $g(D) \otimes_{\mathbb{R}} \mathbb{C}$ de $g(D)$ comme un sous-espace de $H(D,E)$. Pour tout $b \in E$, considérons

$$Y_b = \frac{1}{2}(X_b - iX_{ib})$$

et

$$Z_b = \frac{1}{2}(X_b + iX_{ib})$$

Alors

$$X_b = Y_b + Z_b \quad ; \; Y_b(0) = b, \; Z_b(0) = 0.$$

Y_b est la partie $\mathbb{C}$-linéaire de X_b, Z_b la partie $\mathbb{C}$-antilinéaire.

En s'inspirant de E. Cartan [2] (voir aussi [18]), on montre le lemme suivant

Lemme 3.2.1. $\forall b \in E$, $\forall c \in E$,

$$[Y_b, Y_c] \equiv 0$$

Le lemme 3.2.1 signifie que la forme différentielle définie ψ sur un voisinage U de 0 dans D

$$x \stackrel{\psi}{\longmapsto} \psi(x) = \{b \longmapsto Y_b(x)\}^{-1}$$

est fermée. D'après le lemme de Poincaré, il existe une application holomorphe $f : U \longrightarrow E$ telle que $df = \psi$ et que $f(0) = 0$. Comme $f'(0) = id$, f est une carte locale de D qui envoie un voisinage V de 0 dans D sur un voisinage W de 0 dans E. La carte locale f que nous venons de définir vérifie les propriétés suivantes.

<u>Proposition 3.2.2.</u> Dans la carte locale f que nous venons de définir, $Y_b \equiv b$. On en déduit

(i) le groupe $G_o(D)$ est linéaire;

(ii) l'algèbre de Lie $\mathfrak{g}(D)^+$ est linéaire;

(iii) $x \longmapsto Z_b(x)$ est un polynôme homogéne de degré 2.

Nous noterons $Z(b,x,x)$ l'application trilinéaire associé, $\mathbb{C}$-linéaire symétrique en les deux dernières variables, $\mathbb{C}$-antilinéaire en la première variable.

Ainsi, nous avons construit une carte locale dans laquelle le groupe $G_o(D)$ et l'algèbre de Lie $\mathfrak{g}(D)$ vérifie des propriétés semblables à celles de l'exemple 3.1.6. En fait, la carte f se prolonge comme le montre le théorème suivant.

<u>Théorème 3.2.3.</u> Soit D un domaine borné symétrique. La carte locale f que nous avons définie se prolonge en un isomorphisme de D sur un domaine cerclé étoilé Δ. En particulier, D est contractile et simplement connexe.

<u>Idée de la démonstration.</u> On considère le groupe à un paramètre défini au voisinage de 0 dans D, à l'aide de la carte f

$$(\theta,x) \longmapsto \sigma_\theta(x) = f^{-1}(e^{i\theta}f(x)).$$

On montre que ce groupe agit par automorphisme intérieur sur l'algèbre de Lie $\mathfrak{g}(D)$, et, si on suppose que D est simplement connexe, on peut le faire agir sur le groupe $G(D)$, et ainsi, on le prolonge en un groupe d'automorphismes analytiques de D que nous noterons encore σ_θ. On définit le prolongement F de f par la formule

$$F(x) = \frac{1}{2\pi}\int_o^{2\pi} \sigma_\theta(x)\, e^{-i\theta} d\theta.$$

Le fait que D est étoilé provient du fait que D est complet pour la distance de Carathéodory (voir chapitre 6) et que c'est donc un domaine d'holomorphie. La démonstration (voir [18] pour les détails) est en fait un peu plus compliqué parce que, on ne sait pas, a priori, que D est simplement connexe. Ceci oblige, dans un premier temps, à prolonger le groupe σ_θ en un groupe à un paramètre d'automorphismes de $\tilde{D}$, le revêtement universel de D.

D'après l'exercice 1.2.7., la réalisation de D comme un domaine cerclé borné, est unique, à un isomorphisme linéaire prés. Nous supposerons donc que D est un domaine borné cerclé symétrique. Le groupe d'isotropie de l'origine $G_o(D)$ est un sous-groupe du groupe linéaire et l'application trilinéaire Z permet de le caractériser complètement.

<u>Théorème 3.2.4.</u> Soit D un domaine borné cerclé symétrique d'un espace de Banach complexe E. Soit $f \in GL(E)$. Pour que $f \in G_o(D)$, il faut et il suffit que $\forall b \in E$, $\forall x \in E$,

$$f(Z(b,x,x)) - Z(f(b),f(x),f(x)) = 0$$

La démonstration est, dans ses grandes lignes, semblables à la démonstration du théorème précédent. On considère f comme défini dans un voisinage de 0 dans D. La condition précédente montre que f agit par automorphisme intérieur sur l'algèbre de Lie $\mathfrak{g}_1(D)$ engendré par $\overline{\mathfrak{g}(D)}$. Elle agit sur le groupe de Lie $G_1(D)$ associé, et ceci suffit pour "prolonger" f en un automorphisme analytique de D qui est, bien sûr, egal à f.

Rappelons le théorème suivant, dû à Harris et Kaup [10].

Théorème 3.2.5. Soit A une algèbre de Banach réelle avec unité. Soit G un sous-groupe du groupe multiplicatif $A^\star$ de A. Supposons qu'il existe un entier N et une famille $(P_i)_{i \in I}$ de polynômes de degré $\leq N$ sur A tels que

$$G = \{f \in A^\star \mid P_i(f) = 0, \forall i \in I\}.$$

Alors G est un sous-groupe de Lie réel de $A^\star$.

Du théorème 3.2.5 et du théorème 3.2.4, on déduit que $G_o(D)$ est un sous-groupe de Lie réel de $GL(E)$. Ceci entraîne le théorème suivant.

Théorème 3.2.6. Soit D un domaine borné symétrique. Alors le groupe $G(D)$ a une structure de groupe de Lie réel compatible avec sa topologie et telle que l'application

$$G(D) \times D \longrightarrow D$$

$$(g,x) \longmapsto g(x)$$

soit analytique réelle.

On déduit facilement de notre étude les deux corollaires suivants.

Corollaire 3.2.7. Pour tout $a \in D$, il existe un voisinage U de a dans D et une application analytique réelle

$$U \longrightarrow G(D)$$

$$b \longmapsto f_b$$

telle que $f_b(a) = b$.

Corollaire 3.2.8. L'application

$$D \longrightarrow G(D)$$

$$a \longmapsto s_a$$

est analytique réelle.

3.3. Domaines bornés symétriques et systèmes triples de Jordan.

Nous avons le théorème suivant [12]

Théorème 3.3.1. Soit D un domaine borné symétrique d'un espace de Banach complexe E. L'application trilinéaire Z associée vérifie les deux propriétés suivantes

(i) Il existe une norme equivalente sur E tel que, pour tout $t \in \mathbb{R}$, pour tout $\xi \in E$

$$x \longmapsto \exp(itZ(\xi,\xi,x))$$

soit une isométrie de E.

(ii) pour tous $\xi,\eta,\zeta,x \in E$, on a:

$$2Z(\eta,Z(\xi,\zeta,x),x) - Z(\xi,\zeta,Z(\eta,x,x)) = Z(Z(\zeta,\xi,\eta),x,x)$$

La formule (i) provient du fait que $x \longmapsto iZ(\xi,\xi,x)$ est une rotation infinitésimale de D. Il suffit alors de munir E de la métrique infinitésimale de Carathéodory $\gamma_D(0,\cdot)$. La formule (ii) provient du calcul du crochet $[Z(\eta,x,x),Z(\xi,\zeta,x)]$.

La donnée de (E,Z) vérifiant les propriétés (i) et (ii) ci-dessus s'appelle un système triple de Jordan.

Maintenant, si on considère un système triple de Jordan (E,Z), on peut seulement lui associer une variété normée symétrique (voir Kaup [12]). Pour que cette variété soit isomorphe à un domaine borné, il faut supposer une condition supplémentaire sur (E,Z) (voir par exemple [19]). Cependant, dans ce cas, il est très facile de retrouver le domaine symétrique D à partir de (E,Z). En effet, on a le résultat suivant ([19] et [21]).

Théorème 3.3.2. Soit (E,Z) un système triple de Jordan auquel est associé un domaine borné symétrique D. Alors D est exactement la composante connexe contenant l'origine de l'ensemble

$$\{x \in E \mid id + Z(\cdot,x,x) \in \mathrm{Isom}_{\mathbb{R}}(E)\}.$$

CHAPITRE IV

Automorphismes analytiques des produits continus de domaines bornés et domaines bornés symétriques irréductibles.

Si on considère le produit $D = D_1 \times D_2$ de deux domaines bornés D_1 et D_2, une question naturelle est de calculer les automorphismes analytiques de D. Bien sûr, il y a toujours les automorphismes de la forme

$$f(x,y) = (f_1(x), f_2(y)),$$

où f_1 est un automorphisme de D_1, et f_2 un automorphisme de D_2. En fait, ce sont presque les seuls, comme le montre le théorème suivant, dû à H. Cartan [7].

<u>Théorème</u>. Soient $D_1 \subset \mathbb{C}^{n_1}$ et $D_2 \subset \mathbb{C}^{n_2}$ deux domaines bornés. Alors tout automorphisme f de $D = D_1 \times D_2$, suffisamment proche de la transformation identique, s'ecrit

$$f(x,y) = (f_1(x), f_2(y)),$$

où f_1 est un automorphisme de D_1, et f_2 un automorphisme de D_2.

Ce théorème se généralise facilement au produit de deux domaines bornés d'espaces de Banach complexes. Le théorème obtenu n'est pas suffisant, car il ne permet pas de traiter des produits infinis, comme, par exemple, le boule-unité ouverte de $l^\infty(\mathbb{N})$. Nous allons même montrer que l'on peut traiter des "produits continus", comme par exemple, la boule-unité ouverte B de l'espace $C(S,\mathbb{C})$ des fonctions continues sur un espace topologique compact S. Nous montrerons que, pour tout automorphisme φ de B suffisamment proche de l'identité, il existe une famille $(\varphi_s)_{s \in S}$ d'automorphismes du disque-unité Δ tel que, $\forall f \in B$,

$$[\varphi(f)](s) = \varphi_s(f(s)).$$

Bien sûr, notre résultat sera valable dans un cadre plus général que nous préciserons.

Ces résultats permettent de donner une définition des domaines bornés symétriques irréductibles dans un espace de Banach complexe, et nous trouverons une espèce de décomposition d'un domaine borné symétrique en "produit continu" de domaines bornés symétriques irreductibles.

4.1. Définitions et résultat fondamental.

Définition 4.1.1. On dit qu'un quadruple $(\mathcal{E},S,p,q)$ où $\mathcal{E}$ et S sont des espaces topologiques, $p : \mathcal{E} \longrightarrow S$ et $q : \mathcal{E} \longrightarrow \mathbb{R}^+$ sont des applications continues, est un espace de Banach au-dessus de S si $(\mathcal{E},S,p,q)$ vérifie les propriétés suivantes:

(i) S est un espace topologique complétement régulier

(ii) pour tout $s \in S$, $\mathcal{E}_s = p^{-1}(s)$ est muni d'une structure d'espace de Banach complexe, avec $q_s = q/\mathcal{E}_s$ comme norme;

(iii) les applications

$$\mathcal{E} \times_S \mathcal{E} \longrightarrow \mathcal{E} \quad \text{et} \quad \mathbb{C} \times \mathcal{E} \longrightarrow \mathcal{E}$$

$$(x,y) \longmapsto x + y \quad (\lambda,x) \longmapsto \lambda x$$

sont continues;

(iv) pour toute section f de $\mathcal{E} \xrightarrow{p} S$, soit $\|f\| = \sup_{s \in S} q(f(s))$.

Soit $\Gamma(S,\mathcal{E})$ l'espace de Banach des sections continues bornées par $\|\cdot\|$ de $\mathcal{E} \longrightarrow S$, muni de la norme $\|\cdot\|$. Nous supposerons que, pour tout $s_o \in S$,

$$\Gamma(S,\mathcal{E}) \longrightarrow \mathcal{E}_{s_o}$$

$$f \longmapsto f(s_o)$$

est surjective.

Nous allons maintenant définir quand un ouvert $B \subset \Gamma(S,\&)$ est un produit continu de domaines $B_s \subset \&_s$.

Soit donc $(\&,S,p,q)$ un espace de Banach au dessus de S, et soit B un domaine borné de $\Gamma(S,\&)$. Pour tout $s \in S$, soit $B_s \subset \&_s$ l'image de B par l'application φ_s

$$\Gamma(S,\&) \xrightarrow{\varphi_s} \&_s$$

$$f \longmapsto f(s).$$

D'après le théorème de Banach, l'application φ_s qui est surjective, est ouverte. Ainsi, B_s est un ouvert borné de $\&_s$.

<u>Définition 4.1.2.</u> On dit que B est produit continu des B_s si B est égal à l'intérieur de l'ensemble

$$\{f \in \Gamma(S,\&) \mid f(s) \in B_s, \forall s \in S\}.$$

<u>Remarque 4.1.3.</u> A tout domaine borné B, est associé une famille de ouverts B_s, et on peut considerer le domaine B' produit continu des B_s.

Le probléme qui nous intéresse est l'étude du groupe des automorphismes analytiques d'un domaine borné B, produit continu de B_s. Commençons par le résultat direct.

<u>Proposition 4.1.4.</u> Soit B un domaine borné produit continu de B_s. Soit $\varphi_s : B_s \longrightarrow B_s$ une famille d'automorphismes analytiques de B_s qui vérifie la condition suivante: pour tout $f \in B$,

$$s \longmapsto \varphi_s(f(s)) \quad \text{et} \quad s \longmapsto \varphi_s^{-1}(f(s))$$

sont des sections continues de $\&$ et appartiennent à B.

Alors

$$f \longmapsto \varphi(f) = \{s \longmapsto \varphi_s(f(s))\}$$

est un automorphisme analytique de B.

La réciproque de cette proposition sera le résultat essentiel de ce chapitre.

Théorème 4.1.5. Soit $(\mathcal{E},S,p,q)$ un espace de Banach au-dessus de S, et soit B un domaine borné de $\Gamma(S,\mathcal{E})$, produit continu de $B_s \subset \mathcal{E}_s$. Supposons que l'une des deux conditions suivantes soit vérifiée:

(1) S est discret.

(2) (a) B est la boule-unité ouverte de $\Gamma(S,\mathcal{E})$;

(b) pour tout $s \in S$, $B \longrightarrow B_{s_o}$ admet une section analytique.

$$f \longmapsto f(s_o)$$

Alors, il existe un voisinage V de l'identité dans G(B) tel que, $\forall \varphi \in V$, il existe une famille $\varphi_s : B_s \longrightarrow B_s$ telle que

$$[\varphi(f)](s) = \varphi_s(f(s)).$$

En fait, d'après [20], on peut remplacer la condition (2) par une condition moins forte.

4.2. Trois lemmes.

Lemme 4.2.1. (H. Cartan [7]). Soit D un domaine borné d'un espace de Banach complexe E. Soit a un point de D, et soit $0 \leq r \leq R$ deux nombres réels tels que $B(a,r) \subset D \subset B(a,R)$. Alors il existe une constante ρ (qui ne dépend que de r et R) telle que, quand le point x de D parcourt un segment de droite d'origine a, la distance de Carathéodory $C_D(a,x)$ est une fonction strictement croissante de $\|x - a\|$ sur $[0,\rho]$.

L'idée générale des deux lemmes qui vont suivre est que, si B est produit des B_s, alors la distance de Carathéodory $C_B(f,g)$ doit être proche de $\sup_{s \in S} C_{B_s}(f(s),g(s))$. Ce résultat reste une conjecture (qui est sans doute inexacte d'ailleurs). Cependant, nous avons les deux résultats suivants.

<u>Lemme 4.2.2.</u> Soit $(\mathcal{E},S,p,q)$ un espace de Banach au-dessus de S, et soit B un domaine borné de $\Gamma(S,\mathcal{E})$ contenant la section nulle, produit continu d'ouverts $B_s \subset \mathcal{E}_s$, vérifiant la condition (1) on (2) du théorème 4.1.5. Soit $B(0,r_o)$ une boule de $\Gamma(S,\mathcal{E})$ complètement intérieure à B. Soit T un ouvert non vide de S, et soit $B_o \subset \Gamma(T,\mathcal{E})$ le produit continu sur T des B_s. Alors, pour tout $r < r_o$, il existe une constante $K(r)$ (indépendante de T) telle que $\forall f \in B$, $\|f\| < r$, $\forall g \in B$, $\|g\| < r$ et telle que $\|f-g\|_{S-T} \leq K(r)\|f-g\|_T$, on ait

$$C_B(f,g) = C_{B_o}(f_{|T},g_{|T}).$$

Remarquons que l'on a toujours

$$C_B(f,g) \geq C_{B_o}(f_{|T},g_{|T}).$$

La démonstration de l'égalité se fait en construisant une application holomorphe $\varphi : B_o \to B$ telle que $\varphi(f_{|T}) = f$, $\varphi(g_{|T}) = g$. Par des tecniques semblables, on montre le lemme suivant

<u>Lemme 4.2.3.</u> Soit $B \subset \Gamma(S,\mathcal{E})$ un domaine borné vérifiant les hypothèses du lemme 4.2.2. Soit $s_o \in S$, et soit T un voisinage ouvert de $s_o \in S$. Soit $f \in B$, $\|f\| < \frac{r_o}{4}$, et soit $\xi_o \in B_{s_o} \subset \mathcal{E}_{s_o}$. Alors, il existe $g \in B$, égale à f sur $S - T$, telle que $g(s_o) = \xi_o$ et que

$$C_B(f,g) = C_{B_{s_o}}(f(s_o),g(s_o)) = C_{B_{s_o}}(f(s_o),\xi_o).$$

4.3 Démonstration du théorème 4.1.5.

Soit $B \subset \Gamma(S,\&)$ un domaine borné contenant la section nulle, produit continu de B_s et vérifiant les hypothèses du théorème 4.1.5. Il suffit donc de montrer que, pour tout automorphisme φ de B suffisamment proche de l'identité, pour tout $s_o \in S$, $(\varphi(f))(s_o)$ ne dépend que de la valeur de f au point s_o. Plus précisément, si $f \in B$ est proche de zéro, si h est une section de $\&$ nulle au point s_o, il suffit de montrer que $[\varphi(f + h)](s_o) = [\varphi(f)](s_o)$.
En fait, on peut même supposer que h est nulle sur un voisinage T de s_o. [Le cas général s'en déduit par un passage à la limite].
Soit $\alpha > 0$ suffisamment petit. Soit f proche de 0, $\|f\|_S < \alpha$. Supposons $\|\varphi - \mathrm{id}\|_{B(0,r_o)} < \alpha$. On trouve que

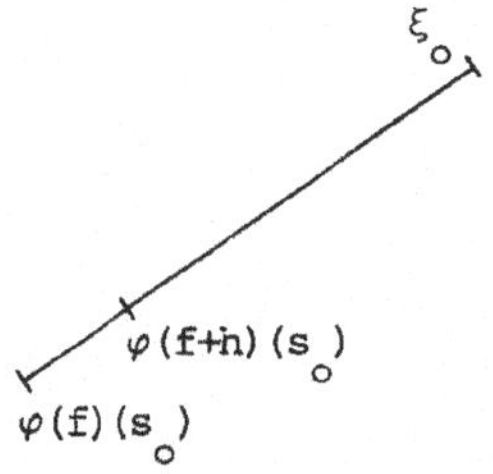

$$\|\varphi(f)(s_o) - \varphi(f + h)(s_o)\| < 2\alpha.$$

Comme $2\alpha < \rho$ (constante introduite au lemme 4.2.1), on peut trouver ξ_o sur une demi-droite passant par $\varphi(f)(s_o)$ et $\varphi(f + f)(s_o)$ et tel que la distance de Carathéodory $C_{B_{s_o}}$ soit strictement croissante sur le segment $[\xi_o, \varphi(f)(s_o)]$.

D'après le lemme 4.2.3, on peut trouver g telle que $g(s_o) = \xi_o$, $g_{|S - T} = \varphi(f + h)_{|S - T}$ et que

(1) $$C_B(\varphi(f + h), g) = C_{B_{s_o}}(\varphi(f + h)(s_o), g(s_o)).$$

Comme la distance de Carathéodory est invariante pour φ et φ^{-1}, on a:

(2) $$C_B(f, \varphi^{-1}(g)) = C_B(\varphi(f), g)$$

(3) $$C_B(f + h, \varphi^{-1}(g)) = C_B(\varphi(f + h), g).$$

Soit B_o le domaine borné de $\Gamma(T,\&)$ produit continu sur T des B_s. On déduit le lemme 4.2.2 que

(4) $$C_B(f, \varphi^{-1}(g)) = C_{B_o}(f_{|T}, \varphi^{-1}(g)_{|T}),$$

(5) $$C_B(f+h,\varphi^{-1}(g)) = C_{B_o}((f+h)_{|T},\varphi^{-1}(g)_{|T}).$$

Cependant, on a:

$$f+h_{|T} = f_{|T}$$

De (4) et (5), on déduit donc

(6) $$C_B(f,\varphi^{-1}(g)) = C_B(f+h,\varphi^{-1}(g)).$$

De (2), (3) et (6), on tire

$$C_B(\varphi(f),g) = C_B(\varphi(f+h),g).$$

En composant avec (1), on trouve

$$C_{B_{s_o}}(\varphi(f)(s_o),\xi_o) \leq C_{B_{s_o}}(\varphi(f+h)(s_o),\xi_o),$$

ce qui, d'après le lemme 4.2.1, suffit à prouver que

$$\varphi(f)(s_o) = \varphi(f+h)(s_o).$$

Le théorème est démontré.

4.4. Domaine borné symétrique irréductible.

Je dirai qu'un domaine borné B de $\Gamma(S,\mathcal{E})$ produit continu de domaines $B_s \subset \mathcal{E}_s$ vérifie la propriété (P) si tout automorphisme analytique φ de B, suffisamment proche de l'identité provient d'une famille $(\varphi_s)_{s\in S}$ d'automorphismes de B_s.

Définition 4.4.1. On dit qu'un domaine borné symétrique D d'un espace de Banach complexe E est réductible s'il existe un espace de Banach $\mathcal{E}$ au - dessus d'un espace topologique compact S, l'ensemble $\{s \in S | \mathcal{E}_s \neq 0\}$ ayant au moins deux éléments, un domaine borné $\Delta \subset \Gamma(S,\mathcal{E})$ produit continu de $\Delta_s \subset \mathcal{E}_s$, vérifiant la propriété (P) et un isomorphisme analytique φ de D sur Δ.

Dans le cas contraire, D est dit irréductible.

Si un domaine borné symétrique D est réductible en produit continu de D_s, on montre facilement que chacun des D_s est symétrique. D'autre part, on déduit facilement des propriétés des domaines bornés symétriques que, si D est réalisé comme un domaine cerclé borné, on peut supposer que φ est induit par un isomorphisme linéaire $T : E \longrightarrow \Gamma(S,\mathcal{E})$ (voir [20] et [22]).

Nous avons déjà vu qu'à tout domaine borné cerclé symétrique D, est associé un système triple de Jordan (E,Z).

<u>Deéfinition 4.4.2.</u> On dit qu'un sous-espace vectoriel complexe I de E est un idéal de Jordan de (E,Z) si $Z(I,E,E) \subset I$ et $Z(E,I,E) \subset I$.

<u>Définition 4.4.3.</u> On dit que D est fortement irréductible si les seuls idéaux de Jordan de (E,Z) sont $\{0\}$ et E.

Il est facile de montrer le théorème suivant

<u>Théorème 4.4.4.</u> Soit D un domaine borné cerclé symétrique fortement irréductible. Alors D est irréductible.

<u>Démonstration.</u> En effet, si D est réductible, il existe un isomorphisme linéaire

$$E \xrightarrow{\sim} \Gamma(S,\mathcal{E}).$$

Soit $s \in S$, $\mathcal{E}_s \neq 0$. Alors

$$I = \{f \in \Gamma(S,\mathcal{E}) \mid f(s) = 0\}$$

est un idéal de Jordan de E, distinct de $\{0\}$ et de E.

On montre d'autre part le résultat suivant

<u>Théorème 4.4.5.</u> Soit D un domaine borné cerclé symétrique de $\mathbb{C}^n$. Alors, les conditions suivantes sont équivalentes.

(i) D est irréductible,

(ii) D est fortement irréductible.

Par un calcul élémentaire [22], on montre le résultat suivant

Théorème 4.4.6. Soit B le boule-unité ouverte d'un espace de Hilbert H, on même plus généralement de $\mathcal{L}(\mathbb{C}^n,H)$. Alors B est fortement irréductible, et par suite, irréductible.

Enfin, on montre le théorème suivant qui montre que, en dimension infinie, les notions d'irreductible et de fortement irréductible ne coincident pas toujours.

Théorème 4.4.7. Soient H et K deux espaces de Hilbert de dimension infinie, et soit B la boule-unité ouverte de $\mathcal{L}(H,K)$. Alors B est un domaine borné cerclé symétrique irréductible et non fortement irréductible.

Pour voir que B n'est pas fortement irréductible, il suffit de remarquer que les opérateurs compacts, par exemple, forment un idéal de Jordan non trivial.

4.5. Décomposition d'un domaine borné cerclé symétrique en produit continu d'irréductibles.

Soit D un domaine borné cerclé symétrique d'un espace de Banach complexe E. Soit (E,Z) le système triple de Jordan associé. Si I est un idéal de Jordan de (E,Z), Z passe au quotient et definit un système triple de Jordan $(E_{/I}, Z_{E/I})$. Le domaine borné symétrique qui lui est associé n'est autre que l'image de D dans $E_{/I}$. On dit que I est irréductible si le domaine borné cerclé symétrique associé à $(E_{/I}, Z_{E/I})$ est irréductible.

Théorème 4.5.1. Soit D un domaine borné cerclé symétrique d'un espace de Banach complexe E. Alors il existe un espace topologique S complétement régulier, un espace de Banach $\mathcal{E}$ au-dessus de S et un isomorphisme linéaire φ de E sur un sous-espace vectoriel fermé de $\Gamma(S,\mathcal{E})$ tels que

(i) pour tout $s \in S$, $\varphi(E)(s) = \mathcal{E}_s$;

(ii) l'application qui, à $s \in S$, associe le sous-espace vectoriel de $\varphi(E)$ des éléments de $\varphi(E)$ nuls au point s, est une bijection de S sur l'ensemble des idéaux fermés irréductibles de E;

(iii) pour tout $s \in S$, $D_s = (\varphi(D))(s)$ est un domaine borné cerclé symétrique irréductible de $\mathcal{E}_s$, il existe un domaine borné cerclé symétrique Δ produit continu des D_s, et $\varphi(D) = \Delta \cap \varphi(E)$.

Ce théorème fournit une espèce de décomposition de D en produit continu de domaines bornés symétriques irréductibles. Pour la démonstration, nous renvoyons le lecteur à [22].

CHAPITRE V

Automorphismes analytiques des domaines cerclés bornés

Nous avons étudié dans les chapitres précédents les automorphismes des domaines bornés symétriques, et nous avons montré qu'ils étaient isomorphes à des domaines cerclés. Nous allons maintenant étudier les domaines cerclés et leurs automorphismes. Nous étudierons en particulier l'orbite de l'origine 0 sous l'action du groupe $G(D)$. Nous étudierons de façon précise la topologie de ce groupe $G(D)$.

5.1. Automorphismes analytiques des domaines cerclés bornés.

Soit D un domaine cerclé borné d'un espace de Banach complexe E. Nous avons déjà vu (exercice 1.2.7) que le groupe d'isotropie de l'origine $G_o(D)$ est linéaire.

Soit $g(D)$ l'algèbre de Lie des transformations infinitésimales de D. En faisant agir s_o (=-id) sur $g(D)$, on obtient une décomposition directe

$$g(D) = g(D)^{+} \oplus g(D)^{-},$$

où $g(D)^{+}$ est l'algèbre de Lie de $G_o(D)$ et est formé d'application linéaires. Comme le crochet de deux éléments de $g(D)^{-}$ appartient à $g(D)^{+}$, on en déduit que $g(D)^{-}$ est formé de polynômes pairs de degré 2. On déduit de la proposition 2.3 que l'application

$$g(D)^{-} \longrightarrow E$$

$$\psi \longmapsto \psi(0)$$

est un isomorphisme de $g(D)^-$ sur un sous-espace vectoriel réel fermé F de E. Du fait que D est cerclé, on déduit que F est un sous-espace vectoriel complexe. Pour tout $\xi \in F$, soit X_ξ l'unique élément de $g(D)^-$ tel que $X_\xi(0) = \xi$. On a:

$$X_\xi(x) = \xi + Z(\xi,x,x)$$

où $Z : F \times E \times E \longrightarrow E$ est une application trilinéaire, $\mathbb{C}$-linéaire symétrique en les deux dernières variables, $\mathbb{C}$-antilinéaire en ξ.

Nous pouvons maintenant énoncer et montrer le théorème suivant

<u>Théorème 5.1.1.</u> (Braun-Kaup-Upmeier). Soit D un domaine cerclé borné d'un espace de Banach complexe E. Alors il existe un sous-espace vectoriel complexe fermé F de E tel que l'orbite de l'origine 0 sous l'action de $G(D)$ soit exactement $D \cap F$; $D \cap F$ est un domaine borné symétrique de F, et $D \cap F$ est le seule orbite qui soit un sous-ensemble analytique complexe de D.

<u>Démonstration</u>. Pour montrer que l'orbite de l'origine est $D \cap F$, le point essentiel est de montrer que $Z(F,F,F) \subset F$. On commence par remarquer que $iZ(\xi,\xi,\cdot)$ est une transformation infinitésimale $\in g(D)^+$, et que, par suite, elle laisse stable F. Soient donc $\xi \in F$, $x \in F$: on remarque que

$$Z(\xi,x,x) = Z(\xi + x,\xi + x,x) - Z(\xi,\xi,x) - Z(x,\xi,x) - Z(x,x,x),$$

ce qui prouve que $Z(\xi,x,x) \in F$.

Montrons maintenant que $D \cap F$ est le seul sous-ensemble analytique complexe qui soit une orbite. Supposons, et on peut se ramener à ce cas, quitte à considérer l'enveloppe d'holomorphie de D que D est étoilé. Soit

$$\Omega = \{x \in D \mid G(D)x \text{ est un sous-ensemble analytique complexe de } D\}.$$

Soit $x \in \Omega$, $x \neq 0$. Alors $G(D)x \cap \mathbb{C}\, x$ est un sous-ensemble analytique complexe d'un disque $\Delta_r \subset \mathbb{C}$ qui contient $|t| = 1$. C'est donc Δ_r tout entier, et $0 \in G(D)x$.

c.q.f.d.

On déduit de ce théorème le corollaire suivant [1] qui améliore le résultat de H. Cartan (voir exercice 1.2.7).

Corollaire 5.1.2. Soient D_1 et D_2 deux domaines cerclés bornés analytiquement isomorphes. Alors il existe un isomorphisme linéaire $f : D_1 \xrightarrow{\sim} D_2$.

Démonstration. Soit $g : D_1 \longrightarrow D_2$ un isomorphisme analytique de D_1 sur D_2. L'orbite de $g(0)$ dans D_2 qui est isomorphe à l'orbite de 0 dans D_1 est donc une sous-variété analytique. Il existe donc $h \in G(D_2)$ tel que $h \circ g(0) = 0$. D'après le résultat de H. Cartan, $f = h \circ g$ est un isomorphisme linéaire de D_1 sur D_2.

5.2. Topologie de la convergence uniforme locale.

Nous allons montrer le théorème suivant qui, à ma connaissance, est nouveau, même en dimension finie.

Théorème 5.2.1. (Vigué-Isidro), Soit D un domaine cerclé borné d'un espace de Banach complexe E. Sur le groupe $G(D)$ des automorphismes analytiques de D, les deux topologies suivantes coincident:

(i) la topologie de la convergence uniforme locale,

(ii) la topologie de la convergence uniforme sur D.

Soit donc D un domaine borné d'un espace de Banach complexe E. Soit F le sous-espace vectoriel complexe fermé de F tel que $G(D)\,0 = D \cap F$. Nous avons alors la proposition suivante

Proposition 5.2.2. Soit M un nombre réel suffisamment grand. Alors il existe une application continue ρ de $D \cap F$ dans l'ensemble des nombres réels > 0 telle que tout automorphisme f de D tel que $f(0) = a$ se prolonge en une application holomorphe de $D_{\rho(a)} = \{x \in E \mid d(x,D) < \rho(a)\}$ dans $B(0,M)$.

Ce résultat (voir aussi [21]) est dû au fait que les transformations infinitésimales sont des polynômes de degré ≤ 2, et on peut donc intégrér l'équation differentielle $\frac{dx}{dt} = \psi(x)$, où $\psi \in g(D)$, un peu à l'exterieur du domaine D.

On montre alors la

<u>Proposition 5.2.3.</u> Soit $(f_n)_{n \in \mathbb{N}}$ une suite d'automorphismes analytiques de D, et supposons que f_n converge vers l'identité, au sens de la convergence uniforme locale. Alors f_n converge vers l'identité uniformement sur D.

<u>Démonstration.</u> On peut écrire

$$f_n = g_n \circ h_n,$$

où $g_n \in GL(E)$ converge vers l'identité, et où $h_n = f_{X_{\xi_n}}(1,\cdot)$ avec $\xi_n \longrightarrow 0$. Le résultat se démontre par un calcul de majorations de solutions d'équations différentielles.

On peut alors montrer le théorème 5.2.1. soit $(f_n)_{n \in \mathbb{N}}$ une suite de $G(D)$ convergeant vers $f_o \in G(D)$ au sens de la convergence uniforme locale. On en déduit que $f_o^{-1} \circ f_n \longrightarrow id$ pour la topologie de la convergence uniforme locale. D'après la proposition 5.2.3, $f_o^{-1} \circ f_n \longrightarrow id$ uniformément sur D. Comme d'après la proposition 5.2.2, f_o est défini sur un voisinage D_ε de D, les inégalités de Cauchy pour la dérivée f_o' de f_o montrent que f_n converge vers f_o uniformément sur D.

Enfin, on déduit de ces considérations le théorème suivant

<u>Théorème 5.2.4.</u> Soit D un domaine cerclé borné. Alors $G(D)$, muni

de la distance de la convergence uniforme sur D, est complet.

Ce théorème prouve que, en général, sur G(D), la structure uniforme de la convergence uniforme locale, et la structure uniforme de la convergence uniforme sur D ne coincident pas.

Signalons aussi que le théorème 5.2.1 devient faux si on ne suppose pas que D est cerclé.

5.3 Automorphismes des domaines bicerclés bornés.

Comme je l'ai déjà dit, on aimerait généraliser un certain nombre de théorèmes démontrés dans le cas des domaines bornés symétriques au cas des domaines cerclés bornés. Pour l'instant, on sait seulement le faire dans le cas des domaines bicerclés bornés que nous allons maintenant définir.

Définition 5.3.1. On dit qu'un domaine borné D d'un espace de Banach complexe E est bicerclé (relativement à une décomposition directe de $E = U \oplus V$) si l'origine 0 appartient à D et si D est stable par les automorphismes linéaires de E

$$(u,v) \longmapsto (e^{i\theta_1}u, e^{i\theta_2}v) \quad (\theta_1 \in \mathbb{R}, \theta_2 \in \mathbb{R})$$

Soit donc D un domaine bicerclé borné d'un espace de Banach complexe E (relativement à une décomposition directe de $E = U \oplus V$). Bien sûr, D est un domaine cerclé, et dans tout ce paragraphe, nous supposerons que le sous-espace F de E défini au théorème 5.1.1 est égal à U. On montre alors le théorème suivant (voir [18] et [1]).

Théorème 5.3.2.

(1) Soit $f \in G_o(D)$. Alors $\forall \xi \in F$, $\forall x \in E$

$$f(Z(\xi,x,x)) = Z(f(\xi),f(x),f(x)).$$

(2) Soit $\psi \in g(D)^+$. Alors $\forall \xi \in F$, $\forall x \in E$,

$\psi(Z(\xi,x,x)) = Z(\psi(\xi),x,x) + 2Z(\xi,\psi(x),x)$.

(3) Soit p une projection telle que $i\,p \in g(D)$. Soit $E_1 = \operatorname{Im} p$, $E_o = \operatorname{Ker} p$, $F_1 = E_1 \cap F$, $F_o = E_o \cap F$. Alors $Z(F_\mu, E_\nu, E_\sigma) \subset E_{-\mu+\nu+\sigma}$ • $(\mu,\nu,\sigma = 0$ ou $1)$.

(4) On déduit de (3) que

$Z(U,U,U) \subset U$, $Z(U,U,V) \subset V$, $Z(U,V,V) = \{0\}$.

Nous allons maintenant caractériser le groupe d'isotropie $G_o(D)$ de l' origine dans $G(D)$.

Théorème 5.3.3.

(1) Soit $f \in G_o(D)$. Alors $f = (f_1, f_2)$ où f_1 est un automorphisme linéaire de U, et f_2 un automorphisme linéaire de V.

(2) Soit $f = (f_1, f_2)$, où $f_1 \in GL(U)$, $f_2 \in GL(V)$. Pour que f appartienne à $G_o(D)$, il faut et il suffit que $f_2 \in G_o(D \cap V)$ et que $\forall \xi \in U$, $\forall x \in E$, on ait

$f(Z(\xi,x,x)) = Z(f(\xi),f(x),f(x))$.

On déduit du théorème 5.3.3 et de [10] que, si V est de dimension finie, $G_o(D)$ a une structure de groupe de Lie réel compatible avec sa topologie. On a donc le théorème suivant.

Théorème 5.3.4. Supposons que V est de dimension finie. Alors le groupe $G(D)$ a une structure de groupe de Lie réel compatible avec sa topologie.

Exercice et exemple 5.3.5. Considerons le cas où $U = \mathbb{C}$. Alors, on peut supposer que $D \cap U$ est le disque-unité $\Delta \subset \mathbb{C}$, et, $\forall \xi \in U$, $\forall x \in U$, on a

$$Z(\xi,x,x) = -\bar{\xi}\, x^2.$$

Supposons que $V = \mathbb{C}$. On peut supposer de même que $D \cap V$ est le disque-unité $\Delta \subset V$ et que

$$Z(\xi,\xi,x) = -r\,\bar{\xi}\,\xi x \in V, \text{ avec } r \in \mathbb{R}^+.$$

On peut montrer [1] que D est égal à l'orbite de $D \cap V$ sous l'action de $G(D)$. Par cette méthode, on retrouve le théorème de Thullen [15] (voir aussi [5]).

<u>Théorème 5.3.6.</u> Les seuls domaines de Reinhardt bornés de $\mathbb{C}^2$ tels que l'origine ne soit pas fixe par les automorphismes analytiques de D sont les suivants:

1) $\Delta \times \Delta = \{(x,y) \in \mathbb{C}^2 \mid |x| < 1, |y| < 1\}$

2) $B_r = \{(x,y) \in \mathbb{C}^2 \mid |x|^2 + |y|^{\frac{1}{r}} < 1\}, \ r > 0.$

Il est facile de voir que $\Delta \times \Delta$ et $B_{\frac{1}{2}} = |x|^2 + |y|^2 < 1\}$ sont homogénes. Dans les autres cas, $F = \{y = 0\}$.

5.4. Les automorphismes de la boule-unité ouverte de quelques espaces de Banach.

Nous allons maintenant appliquer les résultats précédents au calcul des groupes des automorphismes analytiques de la boule-unité ouverte de certains espaces de Banach.

<u>Théorème 5.4.1.</u> Soit D un domaine bicerclé borné d'un espace de Banach complexe E, relativement à une décomposition directe de $E = U \oplus V$. Alors $G(D \cap U)\,0 \supset (G(D)0) \cap U$.

<u>Démonstration.</u> Soit F le sous-espace vectoriel complexe tel que $D \cap F = G(D)\,0$. On montre que

$$Z(U \cap F, U, U) \subset U,$$

ce qui entraîne le théorème.

Le résultat que nous allons montrer maintenant a déjà été montré par E. Vesentini [17] dans un cas un peu plus général. Cependant, notre démonstration est très simple et mérite d'être signalé.

Théorème 5.4.2. Soit p un nombre réel ≥ 1, $p \neq 2$ et $+\infty$. Soit I un ensemble d'indices de cardinal ≥ 2. Soit B la boule-unité ouverte de $l^p(I)$ (l'espace de Banach complexe des suites indexées par I de puissance $p^{\underline{ie}}$ sommables). Alors l'origine 0 de B est invariante par tous les automorphismes analytiques de B.

Démonstration. Supposons que l'orbite $B \cap F$ de 0 sous l'action de G(B) ne soit pas réduite à $\{0\}$. Quitte à faire quelques automorphismes linéaires de B, il est clair qu'il existe une suite $(u_i)_{i \in I} \in F$ avec $u_o \neq 0$. Si $(u_i)_{i \in I}$ appartient à F, il est clair que $(v_i)_{i \in I}$, avec $v_o = u_o$, $v_i = -u_i$, $i \neq 0$ appartient à F. Par suite le vecteur

$$e_o = (1,0,0,\ldots) \in F.$$

Soit U le sous-espace vectoriel complexe engendré par e_o et $e_1 = (0,1,0,\ldots)$. Il existe une décomposition directe $E = U \oplus V$ qui fait de B un domaine bicerclé borné. Alors,

$$B \cap U = \{|x|^p + |y|^p < 1\} \subset \mathbb{C}^2.$$

D'après le théorème 5.4.1, l'orbite $G(B \cap U)\, 0$ est non réduite à $\{0\}$. Or $B \cap U$ n'est pas de la forme annoncée au théorème 5.3.6. Contradiction le théorème est démontré.

5.5. Les domaines de Reinhardt bornés d'un espace de Banach à base.

Les techniques que nous avons développées s'appliquent aussi aux domaines de Reinhardt bornés d'un espace de Banach à base. Commençons par donner une définition.

<u>Définition 5.5.1.</u> Soit $(E,(e_n)_{n \in \mathbb{N}})$ un espace de Banach complexe à base muni d'une base inconditionnelle $(e_n)_{n \in \mathbb{N}}$. Alors tout vecteur $x \in E$ s'écrit de manière unique $x = \sum_{n \in \mathbb{N}} x_n e_n$, et on note $x = (x_n)_{n \in \mathbb{N}}$. On dit qu'un domaine $D \subset E$ est un domaine de Reinhardt si l'origine $0 \in D$, et si, pour tout entier n_o, D est stable par le groupe à un paramètre

$$(\theta,(x_n)_{n \in \mathbb{N}}) \longrightarrow (x_o,\ldots,x_{n_o} e^{i\theta},x_{n_o+1},\ldots).$$

Pour toute partie I de $\mathbb{N}$, soit E_I le sous-espace vectoriel complexe fermé engendré par les $(e_n)_{n \in I}$. Si D est un domaine de Reinhardt borné d'un espace de Banach à base, il est facile de voir que D est cerclé et que D est bicerclé, relativement à toute décomposition directe

$$E = E_I \oplus E_{\mathbb{N}-I}.$$

On montre facilement le théorème suivant.

<u>Théorème 5.5.2.</u> Soit D un domaine de Reinhardt borné d'un espace de Banach à base $(E,(e_n)_{n \in \mathbb{N}})$. Alors il existe une partie I de $\mathbb{N}$ tel que l'orbite de l'origine 0 de E sous l'action de G(D) soit exactement $D \cap E_I$.

On en déduit le corollaire suivant

<u>Corollaire 5.5.3.</u> Si $\mathbb{N} - I$ est fini, la groupe G(D) a une structure de groupe de Lie réel compatible avec sa topologie.

Si D est un domaine de Reinhardt borné d'un espace de Banach à base $(E,(e_n)_{n \in \mathbb{N}})$, on peut étudier avec soin l'application $Z : E_I \times E \times E \longrightarrow E$ associé. Ainsi, on montre le théorème de classification suivant.

<u>Théorème 5.5.4.</u> Soit D un domaine de Reinhardt borné homogène d'un espace de Banach à base $(E,(e_n)_{n \in \mathbb{N}})$. Alors il existe une partition de $\mathbb{N}$ en une réunion de sous-ensembles $(I_p)_{p \in P}$. Pour chaque

$p \in P$, E_{I_p} admet une norme hilbertienne, compatible avec la norme donnée et telle que $D \cap E_{I_p}$ soit la boule-unité ouverte de E_{I_p}. L'espace E est isomorphe à $\prod_{p \in P}^{\circ} l^2(I_p)$ des suites indexées par P d'éléments de E_{I_p} tendant vers 0 quand p tend vers l'infini, et D est l'intersection du produit des $D \cap E_{I_p}$ avec E.

CHAPITRE VI

Automorphismes analytiques des domaines bornés et distances invariantes.

L'étude des automorphismes des domaines bornés utilise de façon fondamentale les distances invariantes. Nous allons montrer ici quelques applications des distances invariantes dans notre étude et aussi quelques autres résultats qui me semblent intéressants. Pour la définition et les propriétés de la distance de Carathéodory et de la distance intégrée de Carathéodory, nous renvoyons le lecteur au livre de Franzoni-Vesentini [8].

6.1. Domaines bornés homogènes et domaines complets pour la distance de Carathéodory.

Nous avons utilisé au chapitre 3 le résultat suivant

Théorème 6.1.1. Soit D un domaine borné homogène d'un espace de Banach complexe E. Alors D est complet pour la distance de Carathéodory C_D.

Démonstration. Soit a un point de D. Il existe un nombre réel $r > 0$ tel que la boule

$$B_C(a,r) = \{x \in D \mid C_D(a,x) < r\}$$

soit complétement interieure à D. Soit $(x_n)_{n \in \mathbb{N}}$ une suite de Cauchy pour C_D. Il existe donc $n_o \in \mathbb{N}$, tel que, pour $n \geq n_o$, $C_D(x_n,x_{n_o}) < r$.

Soit f un automorphisme analytique de D tel que $f(x_{n_o}) = a$. Alors $(f(x_n))$ est une suite de Cauchy pour C_D contenue dans $B_C(a,r)$ pour n assez grand. Elle converge donc vers $b \in B_C(a,r)$, ce qui prouve que x_n converge vers $f^{-1}(b)$.

On peut aussi, à l'aide de la distance de Carathéodory C_D, améliorer les hypothèses de certains théorèmes. Ainsi, nous avons l'exercice suivant, que je laisse au lecteur.

Exercice 6.1.2. Soit D un domaine borné, et supposons que D est symétrique par rapport à tous les points d'un ouvert U non vide $\subset D$. Montrer que D est homogène et donc symétrique.

6.2. Isométries pour la distance de Carathéodory.

On sait si $f : D_1 \longrightarrow D_2$ est un isomorphisme analytique de D_1 sur D_2, alors f est une isométrie pour la distance de Carathéodory. On peut se demander si réciproquement, les isométries pour la distance de Carathéodory sont des isomorphismes analytiques. Une première réponse à cette question est apportée par le théorème suivant.

Théorème 6.2.1. (Harris-Vigué [11]). Soient $D_1 \subset E_1$ et $D_2 \subset E_2$ deux domaines bornés d'un espace de Banach complexe. Supposons que D_1 soit complet pour C_{D_1}. Soit $f : D_1 \longrightarrow D_2$ une application holomorphe telle que f soit une isométrie pour la distance de Carathéodory et qu'il existe $x \in D_1$ tel que $f'(x)$ soit un isomorphisme linéaire de E_1 sur E_2. Alors f est un isomorphisme analytique de D_1 sur D_2.

Remarquons que, dans le cas de la dimension finie, l'hypothèse que $f'(x)$ est un isomorphisme linéaire de E_1 sur E_2 peut être omise. L'hypothèse que D_1 est complet pour C_{D_1} est essentielle, même dans le cas des applications d'un domaine borné D dans lui-même, comme le montre l'exemple suivant.

Exemple 6.2.2. Soit P le demi-plan de Poincaré $\subset \mathbb{C}$, et soit $D = P \setminus \bigcup_{n \in \mathbb{N}} \{i - n\}$. Soit $f : D \longrightarrow D$ défini par

$$f(z) = z + 1.$$

Alors f est une isométrie pour C_D, mais n'est pas un automorphisme analytique de D.

Cependant, en dimension finie, on déduit de H. Cartan [4] le résultat suivant

Théorème 6.2.3. Soit D un domaine borné de $\mathbb{C}^n$. Soit a un point de D, et soit $f : D \longrightarrow D$ une application holomorphe telle que $f(a) = a$. Alors les conditions suivantes sont équivalentes:

(i) $f'(a)$ est une isométrie surjective pour la métrique infinitésimale de Carathéodory $\gamma_D(a,\cdot)$;

(ii) Spec $(f'(a))$ est formé de nombres complexes de module 1;

(iii) le déterminant jacobien de f au point a est de module 1;

(iv) f est un automorphisme analytique de D.

Nous nous sommes intéressés à la généralisation de ce théorème au cas des domaines bornés dans les espaces de Banach complexes. Il est clair que (iii) n'a pas de sens, par contre (i) et (ii) ont un sens. Franzoni et Vesentini [8] ont montré que, même si B est la boule-unité d'un espace de Banach, et a l'origine de E, (ii) n'entraîne pas (iv). Cependant, sous ces hypothèses, on voit facilement que (i) entraîne (iv). Nous allons maintenant construire un domaine cerclé borné D d'un espace de Banach complexe E tel que, dans D, (i) n'entraîne pas (iv). En effet, soit

$$C = \{(x,y) \in \mathbb{C}^2 \mid |x| + |y| + \alpha|xy|^p < 1\} \quad (p \in \mathbb{N}, \alpha > 0)$$

$$A = \{(x,y) \in \mathbb{C}^2 \mid |x| + |y| < 1\}.$$

Soit $\varphi : C \longrightarrow A$ défini par

$$\varphi(x,y) = (x + \alpha(xy)^p, y).$$

Il est clair que $\varphi'(0)$ est une isométrie pour la métrique infinitésimale de Carathéodory et que, si $p \geq 2$ et α est assez grand, φ n'est pas injective. Considerons l'espace de Banach $E \subset \prod_{n \in \mathbb{Z}} (\mathbb{C}^2)_n$ des suites (x_n, y_n) telles que

$$\|(x_n, y_n)\| = \sup_{n \in \mathbb{Z}} (|x_n| + |y_n|) < +\infty,$$

muni de la norme que vous venons de definir.

Soit C_n une infinité de copies de C, A_n une infinité de copies de A, et soit $D \subset E$ l'interieur de $\prod_{n \leq 0} C_n \times \prod_{n > 0} A_n$. Soit $f : D \longrightarrow D$ définie par

$$f((x_n, y_n)) = (X_n, Y_n)$$

où

$$\left.\begin{aligned} X_n &= x_{n-1} \\ Y_n &= y_{n-1} \end{aligned}\right\} \quad \text{si } n \neq 1,$$

$$(X_1, Y_1) = \varphi(x_o, y_o).$$

Nous avons alors le théorème suivant

<u>Théorème 6.2.4.</u> Le domaine borné D est complet pour C_D. L'application f que nous venons de definir est une isometrie pour $\gamma_D(0,\cdot)$. Cependant, f n'est pas un automorphisme analytique de D.

Ainsi, le théorème 6.2.1 ne peut pas être beaucoup amélioré, du moins dans le cas de la dimension infinie.

6.3. Rapport entre C_D et C_D^i.

Soit D un domaine borné de $\mathbb{C}^n$, et soit C_D^i la distance intégrée de Carathéodory sur D. On sait, bien sûr, que $C_D \leq C_D^i$. De nombreux exemples montrent (voir par exemple [8]) que, en général, $C_D < C_D^i$. Cependant, le problème restait ouvert quand on supposait que

D était fortement complet pour C_D (i.e. que les boules pour la distance de Carathéodory C_D etaient relativement compactes dans D). Nous allons construire un contre-exemple dans ce cas. Soit

$$D = \{(x,y) \in \mathbb{C}^2 \mid |x| + |y| < 1, |xy| < \frac{1}{16}\}$$

Alors D est un domaine de Reinhardt, et c'est un polyédre analytique. Par suite, D est fortement complet pour C_D. Nous avons la proposition suivante.

Proposition 6.3.1. Soit x, $\frac{1}{8} < |x| < \frac{1}{4}$.
Alors

$$C_D((0,0),(x,x)) < C_D^i((0,0),(x,x)).$$

La démonstration de cette proposition se fait de la façon suivante: supposons que x est réel > 0. On montre qu'il existe $f \in H(D,\Delta)$ tel que

$$C_D((0,0),(x,x)) = C_\Delta(f(0,0),f(x,x)).$$

On peut supposer que $f(0,0) = 0$ et que f admet le developpement en série de polynômes suivant.

$$\begin{aligned} f(x,y) &= a(x+y) + \sum_{p \geq 2} P_p(x,y) \\ &= a(x+y) + f_2(x,y), \end{aligned}$$

avec a réel > 0.

Lemme 6.3.2. On a: $a < 1$.

Démonstration. Faisons la démonstration par l'absurde en supposant $a = 1$. Soit (u,v) 2 nombres réels positifs tels que $u + v = 1$ et que $(u,v) \in \bar{D}$. Soit $\varphi : \Delta \longrightarrow D$ definie par

$$\varphi(\lambda) = (\lambda u, \lambda v).$$

Alors $f \circ \varphi$ est une application holomorphe de D dans Δ telle que $f \circ \varphi(0) = 0$ $(f \circ \varphi)'(0) = 1$. Par suite $f \circ \varphi(\lambda) = \lambda$, ce qui prouve que $f_{2|\mathbb{C}(u,v) \cap D} = 0$. La fonction f_2 appartient donc à l'idéal $(uy - vx)$. Ceci prouve que $f_2 \equiv 0$. On aurait alors

$$C_D((0,0),(x,x)) \leq C_\Delta(0,2|x|),$$

ce qui est impossible.

Nous pouvons alors terminer la démonstration de la proposition 6.3.1. On sait que

$$\gamma_D((0,0),(u,v)) = |u| + |v|$$

on déduit des hypothèses et du lemme 6.3.2 qu'il existe $r > 0$ et $\varepsilon > 0$ tel que $\forall x$, $|x| < r$, $\forall y$, $|y| < r$, on ait

$$\gamma_D((x,y),(u,v)) \geq \gamma_\Delta(f(x,y),f'(x,y)\cdot(u,v)) + \varepsilon(|u| + |v|)$$

Un calcul simple montre que

$$C_D^i((0,0),(x,x)) > C_D((0,0),(x,x))$$

dès que $|x| > \frac{1}{8}$.

On a donc montré le

Théorème 6.3.3. Le domaine D est fortement complet pour la distance de Carathéodory C_D. Cependant, sur D, C_D et C_D^i ne coincident pas.

BIBLIOGRAPHIE

[1] R. Braun, W. Kaup and H. Upmeier. On the automorphisms of circular and Reinhardt domains in complex Banach spaces. Manuscripta math. 25, 1978, 97-133.

[2] E. Cartan. Sur les domaines bornés homogènes de l'espace de n variables complexes. Abh. Math. Sem. Univ. Hamburg, 11, 1936, 116-162.

[3] H. Cartan. Les fonctions de deux variables complexes et le probléme de la représentation analytique. J. Maths pures et appliquées, 9e série, 10, 1931, 1-114.

[4] H. Cartan. Sur les fonctions de plusieurs variables complexes. L'iteration des transformations intérieures d'un domaine borné. Math. Z., 35, 1932, 760-773.

[5] H. Cartan. Sur les transformations analytiques des domaines cerclés et semi-cerclés bornés. Math. Ann., 106, 1932, 540-576.

[6] H. Cartan. Sur les groupes de transformations analytiques Hermann, Paris, 1935.

[7] H. Cartan. Sur les fonctions de n variables complexes: les transformations du produit topologique de deux domaines bornés. Bull. Soc. math. Fr., 64, 1936, 37-48.

[8] T. Franzoni and E. Vesentini. Holomorphic maps and invariant distances. Mathematical Studies n° 40, North-Holland, Amsterdam, 1980.

[9] L. Harris. Bounded Symmetric homogeneous domains in infinite dimensional spaces. Lectures notes in Maths, 364, Springer-Verlag, Berlin, 1974, 13-40.

[10] L. Harris and W. Kaup. Linear algebraic groups in infinite dimension. Illinois J. of Maths, 21, 1977, 666-674.

[11] L. Harris and J.P. Vigué. A metric condition for equivalence of domains. Atti della Accademia Nazionale dei Lincei, 8$^{\underline{e}}$ serie, 67, 1979, 402-403.

[12] W. Kaup. Algebraic characterization of symmetric complex Banach manifolds. Math. Ann. 228, 1977, 39-64.

[13] W. Kaup. Bounded symmetric domains in complex Hilbert spaces. Symposia mathematica, 26, 1982, 11 21.

[14] I. Piatetsky-Chapiro. On a problem proposed by E. Cartan. Dokl. Akad. Nauk S.S.S.R., 124, 1959, 272-273.

[15] P. Thullen. Die Invarianz des Mittelpünktes von Kreiskörpen. Math. Ann., 104, 1931, 244-259.

[16] H. Upmeier. Über die Automorphismengruppen von Banach mannigfaltigkeiten mit invarianter Metrik. Math. Ann. 223, 1976, 279-288.

[17] E. Vesentini. Variations on a theme of Carathéodory Ann. Scuola Norm. Sup. Pisa, 4$^{\underline{e}}$ serie, 6, 1979, 39-68.

[18] J.P. Vigué. Le groupe des automorphismes analytiques d'un domaine borné d'un espace de Banach complexe. Application aux domaines bornés symétriques. Ann. scient. Ec. Norm. Sup., 4$^{\underline{e}}$ série, 9, 1976, 203-282.

[19] J.P. Vigué. Les domaines bornés symétriques et les systémes triples de Jordan. Math. Ann. 229, 1977, 223-231.

[20] J.P. Vigué. Automorphismes analytiques des produits continus de domaines bornés. Ann. scient. Ec. Norm. Sup., 4$^{\underline{e}}$ série, 11, 1978, 229-246.

[21] J.P. Vigué. Frontière des domaines bornés cerclés homogènes. C.R. Ac. Sc. Paris, 288A, 1979, 657-660.

[22] J.P. Vigué. Sur la décomposition d'un domaine borné symétrique en produit continu de domaines bornés symétriques irréductibles. Ann. scient. Ec. Norm. Sup., 4$^{\underline{e}}$ série, 14, 1981, 453-463.

[23] J.P. Vigué. Les automorphismes analytiques isométriques d'une variété complexe normée. Bull. Soc. math. Fr., 110, 1982, 49-73.

[24] J.P. Vigué. Sur les applications holomorphes isométriques pour la distance de Carathéodory. Ann. Scuola Norm. Sup. Pisa, 4e série, 9, 1982, 255-261.